Without Edwin's hoarding habits, these letters most probably would have been lost years ago. So, I dedicate this book to everyone who, like me, can't throw things away! With this book, I hope to show my grandchildren the importance of keeping things.

Edwin Henning

Edited by Maureen Penrose

MY NORTHERN ADVENTURE

A Teenager's Letters Home from Liverpool to His Mom During WWII

AUSTIN MACAULEY PUBLISHERS™
LONDON · CAMBRIDGE · NEW YORK · SHARJAH

Copyright in Selection and Editorial Material © Maureen Penrose (2019)

The right of Edwin Henning and Maureen Penrose to be identified as the author and editor, respectively, of this work has been asserted in accordance with US Copyrights laws.

No part of this publication may be reproduced, distributed, or transmitted in any form or by any means, including photocopying, recording, or other electronic or mechanical methods, without the prior written permission of the publisher, except in the case of brief quotations embodied in critical reviews and certain other non-commercial uses permitted by copyright law. For permission requests, write to the publisher.

Any person who commits any unauthorized act in relation to this publication may be liable to criminal prosecution and civil claims for damages.

Ordering Information:
Quantity sales: special discounts are available on quantity purchases by corporations, associations, and others. For details, contact the publisher at the address below.

Publisher's cataloguing in publishing data
Henning, Edwin and Penrose, Maureen
My Northern Adventure:
A Teenager's Letters Home from Liverpool to His Mom During WWII

ISBN 9781641828246 (Paperback)
ISBN 9781641828253 (Hardback)
ISBN 9781641828260 (E-Book)

The main category of the book — Biography & Autobiography / Adventurers & Explorers

www.austinmacauley.com/us

First Published (2019)
Austin Macauley Publishers LLC
40 Wall Street, 28th Floor
New York, NY 10005
USA

mail-usa@austinmacauley.com
+1 (646) 5125767

I would like to thank the executors of Edwin's estate who allowed me to access these letters and his diaries. Also, my husband, Martin, who has supported me in the production of this book.

Introduction

When Harold Edwin Earnest William Henning (known as Edwin) died at the age of 92 on 20th January, 2015, little did I know that I would be thrown into a world of bombing, landladies, and rationing. I had known Edwin for over 30 years – he had been a choir master and organ teacher to my husband in the 1960/70s and later became a surrogate Uncle to our children as they grew up. Throughout the years of him visiting us for lunches and teas, he would tell us numerous stories of his life as a child, his evacuation to Liverpool when he was 17 and his later life. When he died, I thought those stories would just be our memories.

As he grew older, he became a hoarder and on his death, I was asked to sort through his possessions; his property was floor to ceiling full of items. During the two-week sort out, I found two shoe boxes and a folder in a darkened corner. On opening one shoe box, I realized I was looking at every letter Edwin had written home to his parents while he had been evacuated by the company he worked for, Union Castle Shipping Company: They moved their Southampton office to Liverpool because of heavy bombing in 1940. For the next few hours, instead of sorting, I was transported, through the letters of a 17-year-old boy, to wartime Liverpool. I later looked at the folder, and found that he had started to write his memoirs. This book is all the letters from the first shoebox (the second box is letters from after he joined the RAF) with added comments from his memoirs.

Edwin was the only child of Martha, and William Henning. He never married, and was a very private gentleman about his life. Up until 19th June 1940, he lived in a rented house on the Millbrook Road in Southampton. His front room had been sublet to the Lloyds Bank and the third bedroom was rented out to a gentleman. His early life was similar to many boys at that time – school, church, choir, and music lessons.

He describes in his writing how 19th June, 1940 became a pivotal date in his life;

German forces were over-running the countries of the Continent Holland, Belgium, and into France. By the beginning of June of 1940, our Allied Forces were pushed to the Channel coast terminating in their dramatic explosion from Dunkirk. This country was under great threat. Air raid warnings and barrage balloons being frequently hoisted raised the fear of imminent invasion by the Germans. On one or two occasions, we visited out Anderson Shelter in the garden during such early night alerts.

On 19th June, my parents decided to join some of our neighbours who voted to use the shelter in the Recreation Ground nearby. That night things became very noisy...gunfire, aircraft – above ground were the Germans landing? Then amid the explosions, it sounded like perhaps a fire-engine racing along Millbrook road... Out of control crashing though railings, and bumping over the grass mound of the shelter. But it terminated in an effect of being blown along inside. A while later two ARP Wardens from a post in the corner of the Rec came to see if we were alright... When the 'All clear' sounded, about at 3:45 a.m. on 20th June, we found two steps displaced. Branches, and leaves of trees, telephone wires strewn in a row, smoke down the road from what had been the White House Garage. The scene was one of destruction. All the semi-detached houses had been seriously damaged. Ours the furthest from the 'white house' had lost its front windows, the front door was up the top of the stairs, the front bedroom door across the other side of the room. Bits of ceiling down, glass, mortar dust, soot was everywhere. It was an exceedingly devastating experience, we all felt severely shaken.

The middle bedroom where I would have been sleeping was in a similar state, and the bed covered with the same debris. The remains of the supper, and tea on the kitchen table had not been excluded from such an unappetizing covering though having had no sleep there was nowhere to lie down, in shock and so disturbed I went for a walk in the direction of the smoking remains of the White House garage and 'Maisonette' next to it had disappeared. The cemetery wall opposite was pitted... We had been bombed!

I continued up Regent's Park Road, found more destruction at the bend of the road. Marshalls' shop near the entrance to the BAT factory was ruined. Turning into South Mill Road, I discovered a crater in a garden of one house *– yet no broken windows! The walk brought me to King George's Avenue. Several people were wandering about. In Westbury Road, a new road off King George's Avenue, there was a house missing. Returning to what had been home all my life, there was to be no breakfast today.*

Getting time to go to work, luckily my bicycle was undamaged and lived up to its name 'Armstrong'. So, I cycled off from scene of devastation on my usual route to the office which because of the night's experience was unusually refreshing.

It was not long after arriving in the office, I began to receive taunts and gibes enquiring if I had been out on the tiles all night! The suit I was wearing had not been taken off since it was put on the previous morning. I was quite unaware of my appearance. It led to a 'Come here, Boy' by the Chief Clerk. I said that I was in an air raid last night. This caused amusement by the staff who no doubt brought that to be a novel excuse for untidiness. Embarrassed, I was nonplussed in my very junior position. I was also astonished that the staff seemed unaware of the night's events. In the inquisition, Mr. Bourne who had cycled from Totton said that he had encountered disruption and destruction in Millbrook around the 'White Horse' garage area. He also found near Totton the timber yard at Redbridge had been on fire. Through the office hierarchy, I had to go before Mr. G N Cox and having established there in fact had been an air raid, incidentally the first on our country, he sent me home 'to help my mother'.

Over the next couple of weeks, they lived with Edwin's Aunt Ena at Hammonds Green. On 1st August, 1940 Edwin and his parents moved to a new bungalow on the unmade road of Northlands Close, Totton. This bungalow, which his parents initially rented, he bought in 1957 and stayed there until his death.

This book charts, as Edwin put it, *My Northern adventure.*

– Maureen Penrose

Bradford Hotel
Liverpool
8th August, 1940

My Dear Mum,

I arrived at Liverpool at 6:25 p.m. and I am staying at the Bradford Hotel temporarily, as arranged by the Company.
Edwin

***

Letter 1

Union-Castle Mailing S.S.Co. Ltd.,
Victualing & Stores Dept. (3rd Floor)
India Buildings
Water Street
Liverpool
Monday, 12th August, 1940

My Dear Mum,

Well, at last, I have arrived at the Offices on Monday morning, the 12th of August, 1940.

On leaving St. Denys, the train proceeded to London, without any hitch, we proceeded across London to Euston, where we boarded out train for Crewe. The route was on the L.M.S. line. We could not tell where we were going through as all the stations are nameless. Rugby was the first with a name, we stopped here at 3:15 p.m. We did not stop again until we reached Crewe, where we changed into the Liverpool train. We passed through a few more nameless stations until we stopped at Runcorn, a dirty looking place. Then we stopped at every station, we came to. Eventually, we became enveloped by two great soot-black walls rising up from the railway to the town, about 70' above. Every 50 yards we passed under tunnels, which probably carried some main roads overhead. Finally, we arrived in a pitch-black station, so dark that no one could only just detect anything.

The Company had booked rooms for the Staff travelling up – Rice and I to be in Room 34, and so on. The arrangements are as follows;

The Company is debited with all meals, etc. while I remain here, at the Bradford Hotel, but I have to find board elsewhere. I am taking immense steps in this direction to enable me to get settled down as soon as possible and to start making musical arrangements.

I woke this morning at seven O'clock, after having a good night's rest. I feel a little stiff after carrying those wretched cases. My lunch tin by the way, had worn its corners through the paper bag long before, we had arrived at Liverpool. We breakfasted, I had ham, toast, and coffee, which was very enjoyable.

I set out for INDIA Buildings, a ten-minute walk from the Hotel. Well, talk about Offices – 9-storey, containing at least 500 offices! (more possibly)

Our Victualing & Stores Dept. are on the 3rd Floor No. 320. Down on the Ground floor, there is an Arcade, with many shops, one in which I am to have Lunch; I am told there is a church, a bank, and practically everything else you can think of in – INDIA BUILDINGS!

I have not yet seen the docks, but that will come later. I want to see Mr. Hume and find out something from him and what is most important, to find comfortable, decent digs in this dirty city – every building which I have seen so far is jet black, not just jet black but JET BLACK.

I am wondering how you are feeling without me, so far away, but I will soon be in a position to settle down up here, if I can. There is half a gale up here blowing from the Atlantic.

I have been typing this to you this morning, as we have four rooms to our Department and they are full of hampers, parcels etc., which I am going through and as I am alone, with our Assistant Super, in the Docks, I can give time to tell you some of my experiences.

Well, it is now about 1 o'clock. At 12 o' clock, I had a hunt around INDIA BUILDINGS, to try and find out Freight Dept. and to find Mr. Hume. Well I have seen him!!

Now, I am excited. I have asked him if he knew of a place for me to stay at. He says he is staying at a lady's place near some park, and that he knows there to be a single room vacant there, so, he will try to get me fixed up there by next Monday. He asked for the number of our Vict. Offices, so as he can come and have a chat with me. He told me that he has been to the Cathedral, and, and, and, and oh dear, – he says he has sat at the Cathedral Organ, and he says 'IT IS THE LARGEST CATHEDRAL ORGAN IN THE WORLD!' He also said that he may know as organist who will take me in his car to the church somewhere here (I don't know the name just yet – but I will soon!).

Mr. Hume has notified me that there will be a recital at a St. Georges Hall (somewhere in Liverpool) next Saturday afternoon, he hopes to go to it. Perhaps, I will be able to go with him. He has told me that I can go to the Rushworth and Dreaper Organ Works, which are Liverpool, and see many old and new instruments, built, and being remodelled.

Then, after the long and interesting conflab, I lost my way, in India Buildings – that wilderness of offices – I was looking for the café. By 12:30 p.m. I found it in the Arcade, it is the Mecca Café. I was alone, but I saw some of the others from other depts, sitting afar off. I will tell you what I had – Roast Mutton and onion Sauce (This was like my Sunday Dinner) Cream Trifle and 2 glasses of water – 1s 6d (to be charged to the company). What about it! – It's fine.

If you write before Friday at:-

Bradford Hotel,
Tithebarn Street,
Liverpool.

You should be able to catch me there, but if Mr. Hume can get me digs, then I will be there by Monday.

I will write again shortly, but if Mr. Hume can get me in his digs, then I will be there by Monday. I must write to Mr. Pickering, Mrs. Reeve, Mr. Torr, Mr. Stamper, Mr. Stevenson, and "Thingamajig" in Coventry, before today or tomorrow are out.

Well goodbye, Mum, for the present, and <u>DON'T WORRY,</u> |Worry gets you nowhere. Be happy, don't overwork. I shall soon be on the organ again. I am looking forward to seeing Mr. Hume *to hear what his good lady can do, and what he can do for me. (Excuse all this bad typing, but I want to let some steam off).*

This ends my first letter of my Liverpool adventure. I wonder how many I will write to you.

Your affectionate son,

P.S. *I have been to S. GEORGE'S HALL, this is like the Guildhall of Liverpool, strangely enough it was the first building I saw, when leaving Lime Street last Sunday. The organ is placed on the stage and occupies quite two-third of it. The platform rises up in steps towards the sides. Well the organ. Oh Gosh!!! On the first program, I was nearly removed bodily (and spiritually) from the chair. At the close of those 3 pieces, Mr. Ellingford brought on the whole 157 stops.*

* * *

Letter 2

14th–18th August, 1940.

My Dear Mum,

Wednesday evening found me dodging round the streets looking for the Cathedral. This was the 3rd time, I had made any effort to reach the Cathedral.

When I reached Leeds Road, I came to a church, surrounded by about 20 stone steps, in a mound to itself between other roads. Having climbed up the steps I entered – the first church I have visited in Liverpool. It was dedicated to St. Luke. Inside, it is as big as Holy Trinity. The Chancel is one-third, the length of the Church. The organ is divided to two sections, one on the left and the other on the right. The dark oak cases being very elaborately ornamented with parts picked out in gold. It is the left-hand section of the organ where the console is, but, it is locked with wooden doors. I could only see the pedal board and nine-foot combination pedals. According to the number of combinations, I gather it must be a decent organ. But, I will see inside it when I meet the organist one Sunday.

From where I saw a tall square tower some distance away with scaffolding round and two deric cranes on top about 200 feet high. When you see that, you know that, that is the Cathedral. So off I trot in that direction – through Renshaw street, and then into Rodney Street and then... The Cathedral is by no means complete, there being the Chancel, North and South Transepts of immense size, but no nave. Owing to it being rocky around this district, the Cathedral is being built of these reddish-brown blocks of rock. It was rather dark as I entered. As you enter the door, you come in on the level of the pulpit and lectern, more or less at the South transept. Looking up, at a great height, I saw pipes of the organ issuing from the North transept, I have marked my position with a cross x as I describe this to you. I became excited when I saw lights on the organ above the choir screen and I also saw lights shining from inside the organ, and heard sounds, unlike those usually produced by a cathedral organ – it was the sound of a hammer. It echoed and re-echoed. It must be glorious in this great building to hear the organ. No music was done while I was there, but I fancied that the makers were doing something to one of the 32-foot stops, for I heard sounds like guns (you know that C below the lowest C on the piano) – Boooooooom, I saw the chairs and behind that the other console with a light on inside and someone twisting some wires round and round. I am planning to go there again tonight, Thursday.

The two red dots denote that under the organ pipes, here are two arches leading to passages running behind the choir screen, and under the organ. The building, so far is very massive. Thursday lunchtime, I went into the church at the bottom of Tithebarn Street. Similar to All Saints but smaller organ is at the back in the gallery.

Friday, another dull looking day, found me having lunch again in the Mecca Café, I had sausages and mash, open jam tart, and custard. After this lunch, I took steps to the end of Tithebarn St., to see the Parish Church again. I also went into St. Nicolas Church and met a verger or somebody. Who said I hadn't a tongue in my head? I straight away asked him how I could get in touch with the organist (an FRCO by the way), he said they were just going to have a service and that the deputy would be coming. This deputy will ask the organist, who is now on holiday and let me know next Wednesday.

What a busy day, I seem to be doing things. This evening, I took a water trip to Wallasey, and I fixed up my digs for next Sunday! Wallasey is near Birkenhead, the birthplace of the new Mauretania. Wallasey is very like Bassett, and likewise being in all appearances, select. It seems abound with Churches.

It is now Saturday, dull day again, smoky etc., like London in the City. I brought a newspaper, as advised by Mr. Hume earlier in the week, for the purpose of finding out the organ recitals. There are two at St. George's Hall.

It seems I have stacks of news for you! Well, Saturday night, or rather Sunday morning at 1 o'clock I was awakened by the sound of guns, and – bombs! We had no warning – just bombs. I flew down the six flights of stair-cases at the hotel as the bombs were dropping. One could hear the whistle. Flashes could be seen through the blackout material covering the windows. By the time, I had reached the Billiard Room, under the Ground floor of the hotel, it had all subsided. I stayed there for about ¼ hour. I heard afterwards that the bombs were dropped at the Brunswick Docks, about 2 miles away. The plane was brought down.

I had my last breakfast of bacon and eggs, cereals, toast, and tea in the morning and then set off for the Cathedral at 9:50 a.m., arriving there about 10:20a.m. I seated myself at the rear. It was not long before I got in conversation with one of the wardens, or some other official, and before the end of the conversation, he offered to take me around the Cathedral next Sunday, the 25th inst. He told me a better position to sit in. I moved accordingly. Mr. Hume and his wife came in and sat in the row before me. There were many people. The organ opened on a soft passage as the procession entered from the extreme east end, and slowly proceeded towards the near end of the chancel, turned, and went into their respective places. The organ in the meantime grew to a terrific magnitude, then reduced more rapidly to something like one little Salicional, almost inaudible. The singing was exquisite. There was a favourite anthem 'Jesu, joy of man's desiring' accompanied with the organ. At the close of the service 3, lines of the National Anthem were sung, kneeling, and unaccompanied, (Pianissimo). The effect was angelic. The great organ pealed forth tremendous tomes after the service.

Now keep your seat, don't excite yourself, I am coming to a great crisis. Mr. Hume commented on the service and moved to go. He asked me to come with him to be introduced to GOSS-CUSTARD, the Cathedral organist. It happened and Goss-Custard invited me to sit with him for the recital before the evening service AND to turn pages. I nearly went crazy after that. It was raining here all day. I went to settle in my new digs. I left there at 6:10 p.m. and headed for the Cathedral again at 7:20 p.m.

I met Goss-Custard at the appointed place, and he took me through one passage, unlocked a door, then another passage, then unlocked another door, then you were behind one portion of the organ. He unlocked the final door leading to the console. He collected the music and asked me to follow him. He sat himself at the organ and had me seated on a little folding chair at the side.

Gosh... What a console! 110 stops on one side, and ditto on the other. 5 manuals, pedal board, revertible pistons, 4 wind-pressure indicators, and thumb pistons. To describe the capabilities of the organ is next to impossible.

Well, I think I have at last come to the end of the letter.

From your affectionate son,

P.S. Keep all these letters for reference. I suggest you to number the envelopes so, if I refer back to a letter you can see it straight away.

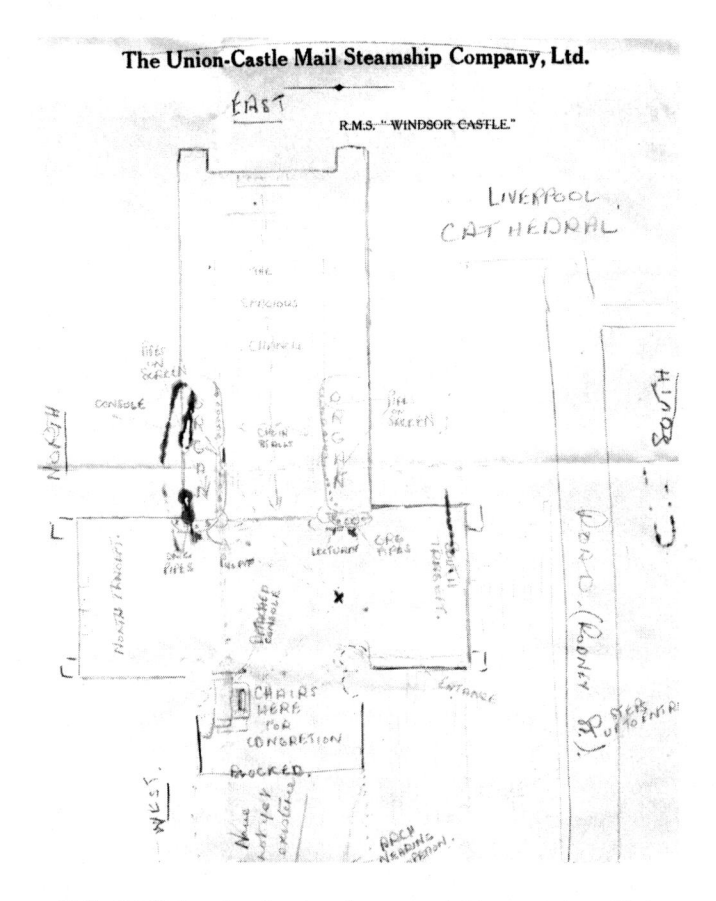

P.S. Well that is what has happened this week in addition to my new address which will be
1 Manville Road
Wallasey
Lancs
Cheerio until I hear from you again. I enclose: 1 shirt, 1 pants, 2 collars, 2 socks, 1 dirty hanky

***

Letter 3

1, Manville Road New Brighton
Wallasey Cheshire
21st August, 1940

My Dear Mum,

I forgot to thank you for your interesting letter, I am sorry but I was so overwhelmed by the happenings up here. I have nothing to report, until lunchtime Wednesday of this week, when I met the Deputy [of St. Nicholas Church, Liverpool], as arranged. She took my address (as above), and afterwards took me up to the organ which is at the back on the level of the gallery. The console protrudes from the instrument like St. Mark's, Southampton.

It is a 3-manual, Radiating Pedal board, draw-stop console, the stops being arranged on stop-jambs facing the performer (like St. Barnabas, Soton). I enclose a memorized copy of the stops, keep it near you as, and when I see it again I will send any necessary amendments to that specification[1].

This deputy may also see an organist who is at Wallasey, and who may allow me to practice over there. The Church is St. Mary's and I do not yet know where it is.

Well considering I have only been here, let me see, 10 days, don't you think I am making good headway, in my blitzkrieg on organs and organists. Only been here 10 days and have seen a Church Organ, with a view to practice. Only been here 7 days, and been introduced to the Cathedral Organist and seen his organ! Dynamite! That's all it is in my music Blitzkrieg. I do not think it will be long before I will be practicing on somebody's organ, if all goes accordingly to plan. I intend this lack of organ playing to be a lull in the storm, but once regaining 'balance' (on the Sw. Pedal) I will drive ahead again with even more determination.

There are several Churches near me at Wallasey, or New Brighton, which ever it is I have seen the outside of St. James,

---

[1] Just like some people collected car or train numbers Edwin collected organ specifications and when we sorted out his bungalow, I remember seeing at least two full books of specifications.

quite a big looking church in grounds to itself, with lawns round, not far from the promenade.

Oh, the wind! It is a gale! A real one, I am not joking. You have never had anything in Southampton like it. You walk, or try to, in one direction and find yourself going to another! No, I am not intoxicated. All the boats in the harbour are rocking like corks, the Atlantic swell lashing around them. Every wave bears a pure white crest of foam. It is treat to be indoors out of it. There has been little or no sun since I have been here, it is all very wintry. But who cares, when I get hold of organ (organs I should have said) and plenty of music: Bach, Adolf Hesse, Gustav Merkel, Edwin Lemare and a whole society of others. To go back to the 16th century Byrd, and so. The things outside are only a means to an end; it blows you there, and blows you back!

Now my landlady has taken my Ration Book, and my identity Card. The former is one page missing, so I am told that it is the meat page. So, would you kindly forward that to me as soon as possible? She gives me to understand that it is with our butcher. The latter is having my change of address marked thereon.

Now I still have some news for you. The company may give me leave after I have been here a month to come home for a weekend. (no doubt to take a review of the damage)

I hope you will keep all these short letters so that at such a time as I come home to stay, I will be able to go over them, and bring back to mind the experiences I had, like a well-written diary. Perhaps, if you were to go over them occasionally, then you will keep a good idea of my position.

From your affectionate son,

Edwin

***

Letter 4

1, Manville Road
New Brighton
Wallasey
Cheshire
21ˢᵗ August, 1940

My Dear Mum,

Reporting from Saturday, I mentioned that I attended two Organ Recitals at S. George's Hall. The afternoon programme included an arrangement of 'Hear my prayer' – Mendelssohn. The treble solo being taken on the solo manual with that mickpathetic Vox Humana and Tremulant.

In the evening programme, the first piece was 'Prelude and Fugue on the name B.A.C.H.! – F Liszt. I have often seen this in programmes of recitals published in the Musical Times, but at last, I have heard it. It is symbolic of Liszt's rhapsodical style. The other six pieces were equally as interesting.

My weekend meals were all included at my digs. My breakfast have been composed of: a plate of corn flakes followed by bacon, tomato, fried bread, toast, butter, jam, and tea. The Sunday lunch was like a feast. Lamb, potatoes, peas, mashed carrots, blackberry tart, rasin [raisin?] tart, custard, and tea. Saturday and Sunday teas at the digs have been quite elaborate. Bread, butter, sardine paste, fish paste, fruit tart, mixed fruit cake, raspberry flavoured chocolate cake, and tea. The weekend suppers have been a number of shredded wheat (brought into the menu owing to my reference of it), cheese biscuits, and cocoa. The item of cocoa appears unheard of in the cafes here.

Sunday morning, I again went to the Cathedral to hear the wonderful choir and organ. The congregation cannot sing the Psalm, as it so was so well expressed by the choir, by the addition of solos, descants, and other modifications to the original chant, that make it more of an anthem. It was quite unique in effect. The Dean is quite a character. His sermons do not fail to arrest the attention of everyone, including the large number of members of the Navy which attended this service. He specializes in poetic and musical history. He based his sermon on the preceding hymn. Yesterdays was on 'Jerusalem'.

Sunday evening, I attended St. James' Church, New Brighton, with my friend (a junite) Mr. Moody, aged 17. Before I had been in the Church 10 minutes, I asked to see the organist. I was escorted to the vestry, where I met Mr. Smith, F.R.C.O., L.R.A.M.

Here, it finishes so suspect other pages has been lost. In his memoirs, he talks about practicing the organ before the end of the month at the Church.

He writes, *"One evening I was practicing and when not playing, I thought the organ had a cipher possibly on a low note and the Pedal Bourdon. I switched the blower off and after some moments expected with the loss of wind supply it would stop. Oh, dear what have I done? A big bang and another one! It was getting dusky and I fled from the church in panic. Down the stone path shrapnel was pinging down. There was an Air Raid. The drone I had heard was by aircraft. So, I sped along in a terrified fright along the Seabank Road when a mobile anti-aircraft gun fired. I reached the digs in Manville Road in quite a state. I probably had not fully recovered from my bomb-out experience in Millbrook. When the landlady brought the evening meal o us the siren went again, and I went a lighter shade of pale.*

She said, "You wouldn't be no good in a war."

So, I told her I had already been bombed out to which she said,

"We had light bombs on Wallasey."...

I, then added, "But you have no air-raid shelters here." To her mind, the thought of such a thing as air raids here, were not even a possibility. My comments implying the possibility evidently led her to thinking in me being a southerner (almost foreign) and so I must be in a league with the Enemy. Where upon, she declared me a Fifth Columnist and would report me to the police. She must have done for the next evening, a Policeman called. There were awkward interviews for me, for the way news was broadcasted, all that was announced would be, "There was an enemy action last night, so many aircraft were shot down, and the loss (if any) of our aircraft over the South Coast or some other area. Mainly, of course, the Battle of Britain was going in the air day and night. The air raid on Millbrook would not be known about unless one lived in that area. So, I told the Policeman I had been bombed out there on the 17th June, and *lived in Totton now before being sent to Liverpool by the Shipping Company. I expressed my concern from the experience the worry I had by the absences of air raid shelters. After a few days, he returned to reassure the Landlady and myself that I was not a Fifth Columnist!*

***

Letter 5

1, Manville Road
New Brighton
Wallasey
Cheshire
31st August, 1940.

My Dear Mum,

I have been very busy at the office during the whole of this week, quite frequently, not leaving the office before seven o'clock in the evening.

On Tuesday, I effected the appointment with Mr. Smith, in the evening whereby arrangements were made for me to receive lessons from him at a reduced fee, as a concession to my agreement to be a member of the choir. That I think, sums up my present musical position.

That evening, I received my first half of a lesson. Wednesday, I had a practice, while he had his boys' choir in the vestry on the opposite. He congratulated me afterwards on my apparent silence during the practice – he did not know that I was practicing as I was so quiet with the organ. He, thereby, wished to know what stops I was using.

Friday, I had another short practice, again with his boys' choir busy in the vestry. After this, I went over to the school opposite, where the full practice was to be held. This is a choir with a few men in it! About a dozen. Various chants, hymns, canticles, and anthems were sung, including an anthem by Edward C. Bairstow.

Thursday night, we had a warning from 11:30 p.m. – 4:00 a.m., the next morning. Incessant A.A. gunfire throughout the raid. Friday at 4 o'clock, the first daylight raid since I have been here. 10:30 p.m. to 3a.m., the next morning, Saturday was the next raid with 'screamers'.

I am still hearing of many lengthy raids in the 'South-coast' area, which I take to be mainly in Southampton, and I trust that things are in a favourable condition.

I am coming home, next weekend.

From your affectionate son,

P.S. I scribbled this while on the pontoon, at the Pierhead.

1.9.40

This place where I am staying is with Mr. Moody and Mr. Hume, he (MR H) did not like his digs so, did not recommend them to me.

Last night, I took a walk along the Promenade where I saw the funfairs and other amusements – I naturally watched others spending their money while I refrained from doing so. My friend Mr. Moody had accompanied me. We returned at 8:20 p.m. , then at 8:30 p.m., the sound of the enemy approaching – a few minutes later, a solo whistle, joined two secs later by a screamer (making a duet), and soon grew to a chorus. The bombs fell after the 15-secs music, which demolished a Town Hall, a warehouse, and a custom House. After that a siren sounded. Several more loads were dropped during the raid, causing a few fires. Every night, this week, we have had lengthy raids leaving only a few hours for sleep.

Keep fit until I try to get down next weekend.

***

Letter 6

1, Manville Road
New Brighton
Wallasey Cheshire
9th September, 1940.

My Dear Mum,

Having left punctually from Southampton Central, we anticipated arriving at Waterloo according to schedule, which caused us to expect to have crawled from the rest of the journey. After stopping at Surbiton, it was gone at 10 o'clock, we pushed off for about two miles, then heard the sound of enemy aircraft overhead. The train braked suddenly, when six bombs were dropped around the train, shaking it violently. We found ourselves under the tables between the compartments, and other places of 'refuge'. The enemy had evidently made a target of our train, but fortunately missed the mark, but brought down some buildings nearby. Glows of fire could be seen in the distance.

The enemy having illuminated the area circled round for three hours, during which period I would have given anything to get a return ticket for Southampton, if the trains had been running. Further bombs were dropped a greater distance away.

It was a great pleasure when the driver took a fit into his head to proceed. We seemed to go about 500 yards then stopped, and the words 'all change' rang out from the porters on the platform. We gathered our belongings and disembarked. We were at Wimbledon. I noticed fragments of glass lying about on the platform where splinters and shrapnel had fallen through the station covering. We were directed to the Air Raid shelter to the west of the station, and by 1:15 a.m., we had settled for the night in this smoke-filled shelter. Passengers from other trains were in the shelter.

Bombs dropped continuously, many far away, with a few near to hand. I actually slept for three-quarters of an hour during the bombardment. By dawn, I went out on to the platform then the sirens went away. I walked along the platform, 50 yards from the shelter and discovered a bomb crater in the middle of the District Line to the left of the platform. Part of the platform had been blown away including part of the roof over this platform. All platforms were covered in glass.

No trains were running to Waterloo. Anyway, by six o'clock, an electric train was coming through from Holborn, we passed through Tooting, Black Friars, and others where I expected to see quite a lot of damage but apparently, the raiders had been more west of London, although around Tooting a number of houses, several factories, and other buildings near the railway were seen to be demolished. We passed through Elephant and Castle but did not see much damage here although two or three [houses] appeared to have been burnt out. On the horizon, great volumes of smoke could be seen rising in the morning air, in the docks area. From Holborn, we bussed to Euston arriving there at 7:40 a.m. Our train for Liverpool left at 8:30 a.m. During the last lap of this eventful journey, I actually went asleep. We arrived at Lime Street at 1:30 p.m. and I went to India Buildings to type this out to you.

That little journey took us 18 hours and 10 minutes! Well I'm here. In the meantime, I remain your distant son, yours affectionately,

* * *

Letter 7

1, Manville Road
New Brighton
Wallasey
Cheshire
13th September, 1940

My Dear Mum,

I thank you very much for your letter, which I think must have been written and dispatched before you received my epistle.

Others of our staff travelling back the successive nights, had equally thrilling experiences in travelling through London. As I could not come back this side of Wimbledon to see what had happened, so did not know the extent of the damage done that night.

I did not do any music that day I arrived, I went to bed at 9 o'clock, and slept peacefully until I heard any 'all clear' some time of the night. The raids at present are reduced in number and length, the main reason being a result that the Air Ministry have taken over the port in my absence on leaving which I hinted might happen when I was with you.

I hope Felix, Christopher, and Peter are still in good spirits. It is a pity that I can't see them more often. I can only imagine an entanglement of string, anchored at one end on the coal-box, platted together and twined around many objects, with Felix and Christopher, both looking very bewildered at the muddle at the other end. Perfect gentleman, they never even murmur at their distress.

As I have found someone who can play Chess up here, I should be very pleased to be in possession of the box of men now somewhere in Totton. I do not know whether it would be worth registering the parcel when you send it, considering the wot-not is only worth 10/6!

I am sending the coat to have the tear put right together with some washing. *Your affectionate son,*

Edwin

***

Letter 8

1, Manville Road
New Brighton
Wallasey
Cheshire
27th September, 1940.

My Dear Mum,

I received your letter of welcome dated 20th inst, on the following Monday 23rd. It is quite a pleasure to have this flow of correspondence. That morning, I received a letter from dear

Mrs. Reeve. It expressed a wish for me to remember her to Mr. Ellingford. She mentioned the re-bombing of the High Street [Southampton], as you did, and also the fact that she did not go to church one Sunday evening. She sent her best wishes for my welfare and safety. On Thursday, I received two more letters. The first was from Mr. Stephenson, he mentions that they carry on with little music at some of the services and that he is trying hard once again to encourage boys in the choir. He advises me to see Chester Cathedral while I am staying here, and also to see the vicar of Seacombe, Mr. Davis, who is an old friend of his. He states that the Dockland Settlement has received only one little bit of damage, and that was caused by shrapnel. The other letter was from Mr. Torr, I regret to say that he has written to say that he is suffering from shock and has been ordered away to Hay in Herefordshire.

Hoping this letter reaches you in time for your birthday, I send my fervent love and best wishes for a happy and safe year to come. I will perform another form of honour to the event when I come down on the 11th October.

Air raids for the past week have been mainly effective in the central Liverpool and docks areas. Early in the week, some warehouses were hit, causing fire and some damage. I have not seen the damage, therefore cannot describe. News from Southampton over the phone regarding this subject deeply grieves me. I have heard with deep regret that considerable damage has been caused to s.

Barnabus Church, bombs dropped in the Inner Avenue and opposite Wadhams. A time bomb dropped near the Central Station and bombs also at Testwood and Bitterne Park. I take that statement to be true as I have heard it confirmed from other sources. In which case, as a reciprocation of theirs when we were distressed in June last. I should like Mr. Pitt and Mr. Hartnel to receive my sympathies especially, as they helped in our distress. I hope the organ is not badly damaged, but do not expect it to be worth much after a bombardment.

Last night, at 7:15 p.m., the sirens went while Mr. Moody and I were taking a walk along the promenade. I suggested the walk, to see an old picturesque café overlooking the sea. It was established in 1595 and is coloured white, with dark beams running at angles. We returned immediately and before reaching our 'base' A.A. guns spoke out. About 5,000 feet up were 5 bombers, I jumped in a Kiosk when they started machine gunning. From there, I saw the gunfire, planes, and everything, although of course, I did not feel very safe. The planes went directly overhead, and one copped it, a flaming onion went through one of them and cut it in two. He caught on fire and fell with sparks coming from the ends of the two portions. He fell over Birkenhead somewhere. The enemy had his revenge during the rest of the raid, the all clear went at 11:00 p.m. Arriving next morning, we saw two dense clouds of smoke rising to right of the blocks of offices near our India Buildings. In the middle of the road, by the building of the right, a bomb had landed blowing out the windows, in that building the windows in Cunard Building, and some in the Royal Liver Building. I should not be surprised if the enemy does not have a go at our India Buildings, and Cunard Buildings, the R.A.C. buildings and the greatest of all the Royal Liver Buildings in future raids. I hope to enclose aerial picture of these buildings. Our offices, as you may gather form part of the business centre of Liverpool, and the picture. I hope to enclose shows the whole of main buildings.

I enclose two other photographs as a part of my expression on your birthday. One is the interior of the Cathedral, and the other is of S. George's Hall. Both buildings still seem to an eyesore to the enemy. The position of the camera to have taken the former picture, must have been in the position in which is it when I go the service. You can see the organ, the pipes of which can be seen on both sides of the church, and the closed console, above the choir on the left.

My victualing midday for the past week has been the following. Cold lunch on Monday, hot on Tuesday, which was grapefruit: Roast mutton, mashed potatoes, carrots: Strawberry ice cream: Coffee. Wednesday, another cold lunch. Friday, again I was served at Maison Lyons, with: Grapefruit: Roast beef, chip potatoes, mashed swedes: Yorkshire pudding: Strawberry ice cream: Coffee. For teas, I have been favouring a little coffee stall by the Pierhead. Hot meat pies, cakes, and tea make quite a filling little meal with expectation of a good supper when I get 'home'. This is usually served by the lady of the house about 8 p.m. or 9 p.m. in the evening. It has been something after the style of the following;

Baked beans on Toast, bread and butter, biscuits, and cocoa. Cold meat, (preserved), salad, bread, and butter. Shredded Wheat and hot milk, sandwiches, biscuits, and Cocoa.

Not to leave out my breakfasts, it invariably commences with cereals, followed by either three skinned sausages, or baked beans on toast, or poached egg on toast, bacon and tomatoes, and bread fried; which is always followed by toast, butter, and marmalade. I have two cups of tea every morning during breakfast.

Hoping this letter finds you in perfect health and safety, I, now conclude this epistle, with a note to the effect that I have not received the Chess set up the present – Friday 27th – but I am not alarmed, as the average parcel takes 6 to 8 days to travel between Southampton and Liverpool.

I am your affectionate son,

P.S. I will forward washing later. No recital owing to Hall being otherwise engaged. Walked through the bombed area of Liverpool. Factories. Warehouses, a pub, three or four air raid shelters, are badly damaged. Several fires over a wide area is still burning. In this district, there are few streets without damage. A huge grocery store is damaged by bomb and fire. Many streets are roped off. It looks a very deplorable spectacle. Could not get on to Southampton by phone yesterday, we put a call on at 9.30 a.m. in the morning but waited in vain all day. Got through this morning and victualing dept. described it as rotten (in the raids).

***

Letter 9

1, Manville Road
New Brighton
Wallasey
Cheshire
4th October, 1940

My Dear Mother,

I was very pleased to receive your parcel with the Chess set and coat enclosed, together with coat hangers. The book has already proved a brightener to my life here, and I hope it will continue to do so until my return next week, my leave is, of course, sensitive to the movements of our fleet.

All I seem to do is to write letters, and prepare them, night after night. LETTERS in the office, then at 'home' – What a life! What an existence! In reading your letter, undated, enclosed in the parcel, I am amused as you were regarding my memories of the cats on the string. The phrase 'easy to haul in when the sirens go' tickles me immensely.

I have received a cheery letter from Auntie Ruth. It must have arrived with your parcel for Mrs. Hart presented me with both as I arrived 'home' Monday evening.

I had a good practice on Tuesday evening without interruption by air raids. I caught the old spirit when playing Mendelssohn's wedding march, and also one of Guilmont's. I also looked at Wagner, Gade, Lemare, and Bach and so as you may gather it formed quite a brilliant stimulus, in these rather dead-end times.

The chess set has already represented many battles. One battle was fought on its arrival, when its possessor lost. Three battles were staged on the following Wednesday, casting over a period from 8:30 p.m. (at the close of supper) until 10:30 p.m. the first battle might have aroused the attention of a spectator, if there had been one there, in the way the pieces entered into conflict far too early in the battle (The pieces, being those fellows on the back lines). This battle concluded unfavourably for its possessor. The second battle was successfully concluded within six moved of the possessor. The third and final battle, waged on for some considerable time, when it was thought that the result would be stalemate. The solution was unfavourable to the possessor.

To turn to the subject of aerial warfare, there is very little to report here. From my observations, the reduction in number of raids is almost staggering. I have not troubled to enumerate the raids as they have passed, but judging from memory, I can say the average has been two a day (spread over 24 hrs) even at night if a warning has gone, enemy aircraft have not approached us more than twice.

I am quite pleased to inform you that every night, including last Sunday, I have been able to go to bed at 10 o'clock, and remain there to dream and slumber, until disturbed by a fierce knock on the door at 7:30 a.m. the next morning. I admit I have gone to bed when a warning has been in progress, but I have not had to go down again as there has been no gun-fire or aircraft about. The reason for this peacefulness may be the result of the very foggy atmosphere we are experiencing just now, in fact, for the whole of this week there has been a yellow fog over Liverpool and district, including the River, of course.

This morning I received your parcel dated, the 2nd October, which is exceptionally quick in comparison with other parcels, I have hitherto received from you. I had just finished and was off to catch the bus for the Seacombe Ferry when it arrived. I hurriedly, extricated the letter so, that I could reply to it during my usual Friday letter to you. I am glad, you liked the card and photographs as I thought you would. Regarding your remark about my being in the docks, you need not worry there as my work does not necessitate my doing so. Although the Ferry boats leave the landing stage just in front of Royal Liver Building, shown on the photograph, and this lading stage is a continuation of the Queen's Landing stage, where I have seen such craft as Duchess of Richmond[2] and others. There is a gate between the two stages preventing one from entering one from the other.

Referring to the tactics of planes, I have known Spitfires to go out early in the evening in the direction of the sea, this was adopted immediately the port was taken over by the ministry.

---

[2] SS Duchess of Richmond was an ocean liner for the Canadian Pacific. During this time, it was transporting evacuated British children to Canada.

I hope to see Mrs. Hart regarding the proposed visit, so that it may be possible for you to return with me when I go back to Liverpool next week. If the raids are as few as they have been this week, it would prove a source of rest for you, and also a great joy to hear the Organs in the Hall, and Cathedral, respectively.

From your letter again, regarding the 'invasion Dream', I say as a headline this week that the Nazis Blitzkrieg has failed, and they have admitted it.

I noted the attached advertisement for St. Barnabus Church, and I am glad to see that they are able to resume the services – but unfortunately without that dear old organ.

I suppose I must close down now, unless I reopen for a P.S. or so, until we meet again next week. At present the pre-calculated date, the 11th of October is my travelling day and I hope to take you back with me on the 14th, Monday night.

From your affectionate son,

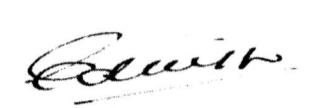

Edwin.

I have seen the proprietor, who has asserted to your occupying the room next to ours (colleague and myself). If you can remember the 6th of Oct. last year we were in Bournemouth. On the 14th we went to Salisbury, when we visited the Cathedral of S. Thomas of Canterbury. Before me, now I have the rough list of stops, I took when leaning over the choir seats to read the stops. I can remember eating Pork pies in the Cloisters and placing a boot lace behind one of the drain pipes there. I wonder of the latter is still there? While in the cloisters, you will also remember that I was studying the Venite in one of the St. Marks Red Book and practicing mentally one of the chants to it. I also thought of the piece 'Cloister Gaith' which as well describes the atmosphere in those precincts.

Hoping to be home by midnight on Friday. I do not want you to bother to go to the stations or anything like that, for I do not know where I shall arrive or even when. One of my shoes is beginning to show signs of wear, due no doubt through walking to St. George Hall and Cathedral so often, and that 3 hours walk a fortnight ago.

***

Letter 10

1, Manville Road
New Brighton
Wallasey
Cheshire

My Dear Mum,

I have just arrived at 10 o'clock at Birkenhead, just an hour behind schedule. Changing at Basingstoke was simple. I slipped down the sub-way to platform number 5. where a G.W.R. train was just backing in. it was not illuminated, but it did not much matter as there was a moon. This was bound for Reading. 'shooting stars' in the distance told of battles over London. Arriving at Reading, I passed through another sub-way to another platform 5 at 11:20 p.m. The train for Birkenhead (as it was through train from here, without further changing) was not to arrive until 1:10 a.m., I parked myself in a waiting room on this platform and fell into conversation with a fellow who is being called up, who had to get to Blackpool. A luggage train passed through at 1:15 a.m., then came an express for Penzance at 1:30 a.m., which blocked the line for twenty minutes while loading mail. By 2:00 a.m., the line was cleared of this train, and the Birkenhead train steamed in and braked. It pushed off for its long journey soon afterwards. The other fellow sat on one side of the compartment, and I the other. I used the length of the seat as a bed, and fell asleep before reaching Oxford. It was 6 o'clock when I awakened, and I was told we were just passed Wolverhampton. I sat up but fell asleep again until I was awakened by the ticket inspector at Shrewsbury. After that little disturbance, I went to sleep again until we had proceeded as far as Chester. It was 8:45 a.m. so, I decided that I had enough sleep for one night and inspired by the bright sunshine, and the fact that I was travelling, I then made my breakfast. I had entered the bag, by the way, before we left Reading last night, and ate up the bread, and butter, sandwiches.

From the train when nearing Birkenhead Ferry across to Liverpool, the water was very troubled and rocked the boat violently. Arriving in Liverpool at 10 o'clock, my repeated predictions regarding the enemy's attraction to the blocks of offices in Liverpool is manifesting itself. I see that the Royal Liver Building has lost most of its windows, but I can see no bomb crater outside so, I take it HE put one inside. The building behind the Mersey Docks and Harbour Board Offices, which is occupied by the R.A.C. had a glancing strike on the further left corner, blowing out the two extreme rooms on the 3rd and the 4th floor, and blowing out all the windows of the 3rd floor, but freakishly leaving all other windows, even directly over the point of contact, and below it, as well as all other windows, intact.

I have been informed, but not yet had time to verify that Wallasey and New Brighton had sustained damage the same night as the bombing of Liverpool on the Saturday night. As before mentioned, Saturday is Hitler's busiest day in Liverpool and district, if we are going to have a bad raid, sure enough, it will come on a Saturday evening or night. News of any other damage has not yet, reached me.

I have made another attack on your sausage rolls and biscuits, as well as, removing the three apples in my pocket (two were eaten during journey from reading to Birk).

I cannot think of anything further to comment upon at present, so will close down now until my usual weekly letter to you, typed on Friday, and P.S. additions appended on Saturday and Sunday.

From your affectionate son,

P.S. Hoping you and dad returned home without undue delay or too many thrilling events.

***

Letter 11

1, Manville Road
New Brighton
Wallasey
Cheshire
18th October, 1940

My Dear Mum,

This is my second letter to you of my 3rd term in Liverpool. (Every term constituting 4 weeks). I know you must be anxiously awaiting this as I am for from you.

I find it difficult to collect sufficient data for this letter as I have only three days on which to report. Regarding the bombing of New Brighton, which I mentioned in my letter of the 15th inst., I allege this to be correct. Mr. Moody confirmed the statement on questioning. He enumerated some of the damage. Bombs were dropped near the library which is only 5 minutes' walk approximately from the above address, but I have not yet identified the damage.

We have received nightly attention by the enemy, the first night, the day of my arrival, he concentrated in the City, the night following, he erased three public shelters from the map, and Thursday night he dealt with a Merseyside town, although it was foggy.

The weather conditions are foggy. The river and banks are covered in a thick blanket, which causes the movement of river traffic to be scarce and that which is in operation, to be very slow, causing unavoidable delay.

One battle on the chess board was in progress last night during the early part of a raid (during which there was heavy gunfire, which was repeated in a succession of 10-minutes deafening roar of continual bombardments). The internal battle at No.1 was prolonged for a whole hour, with the prospect of stalemate until I became sleepy and made a few false steps, which hurriedly concluded the battle.

Having heard nothing regarding raids in the South, I gather that you have had a fairly quiet week and that you have not needed to haul in those lengths of string with the two terrified cats anchored on the table legs.

From your loving son,

Edwin

P.S. Further damage has occurred in Liverpool. Friday night, houses, a garage, and a factory were demolished but not seen by me. Saturday night, of course, the week's night of night, we expected something. As this Sunday, having been to Church this morning and going again this afternoon, I have had no time to see Saturday night's damage. The city, and the town adjacent to New B. received attention. Last night, I actually heard the screamers descending amid the roar of the A.A. guns, but as could be expected I was unable to ascertain the bursting of the bombs with the terrific noise aforementioned barrage. My only clue to their detonation was several slight vibrations through the floor; this will often occur when two or more guns fire together, as is the case here. It is very difficult therefore to discover what is actually happening. This makes the 5th night in succession since my return that a number of eggs have been dropped; eggs ought to be much cheaper!!

* * *

Letter 12

1, Manville Road
New Brighton
Wallasey
Cheshire
25th October, 1940

My Dear Mum,

I, hereby reopen communication with you by dealing with the weather conditions first, by the way of a change from the usual order of procedure. The temperature is not yet down to zero, and is, in fact, nowhere near to it but there is certainly a nip in the air in the mornings and evenings accompanied with mist over the river and its banks. Tuesday, it was I believe, when we had another of those thick, suppressing fogs. It was as dark at 10 a.m. in the morning as a winter day at 4:30 p.m., when black clouds foretelling a fall of snow, hung overhead. This fog formed another course to my breakfast as I crossed the River in Liverpool, it partook in that luminous colour, which tinges every object with a yellow hew. Thus, we were shut in this blanket until noon, when it became less thick, less yellow, and therefore, less suppressing.

As I type this letter my friend, Mr. C.H. Moody, is somewhere in England, seated (I hope), in the Birkenhead train to Banbury. Judging from my own experience of this route, he should be approaching Shrewsbury. I will not see him again until 4th of November, when he will return after spending 9 days in Southampton. I shall not, therefore, be able to have any further battles on the chess board until then. This week, we were engaged in three battles. No. 2 was favourable to me, No.1 was not, and No 3. was favourable to neither of us, watchfulness produced a stalemate.

I have had very little correspondence with which to deal during the past week, other than sending by Registered Post, the sum of 37/6d to Mr. Torr at Hay and enclosing a letter. He states that the vicar of his Parish has given him permission to play on the church organ. He relates to the other Sunday when he gave a recital and goes on to say that it proved a great attraction, for the church was packed to the upmost. He also described the type and position of the instrument.

Perhaps now, we ought to deal with that apparent never-ending subject regarding Air Raids. Up to date, I have heard the number of air raids that Liverpool has been subjected to since their first one in mid-August. That number is 207, of which every evening and night raid, with the exception of one or two, there has been a liberal dropping of bombs.

Last week was War Weapons Week in Liverpool. Posters were displayed at convenient points advertising War Bonds, Saving Schemes, and other attractive propositions. One particular attraction was a display of two enemy planes placed in the forecourt to S. George's Hall. (on the right of the building on your postcard – reminded me when I come down), travelling cinemas were among the propaganda. The proceedings amounted to somewhere in the region of £11,500,000.

Last Sunday, the 20th inst, I attended the Parish Church of St. James, as usual, and it chanced to be their Lifeboat Sunday. In the afternoon, the Bishop of Chester gave the address. An anthem was sung by a fairly large choir, in comparison with Millbrook. There were, incidentally, 10 men and 12 boys. The entire clergy consisted of: the Bishop and his assistant, the vicar, Mr. Goad, and his assistant. Mr. Smith was at the organ. The body of the church was filled to capacity. The Mayor of Wallasey and the Magistrate, and other 'fogies' of the corporation occupied those seats on front right, while members of the New Brighton Lifeboat Institution occupied those on the left.

I am hoping to make beneficial use of the reference and lending libraries in the city in the near future. I viewed the former on Thursday last, with the object of scanning some tomes on my favourite subject, need I mention it – ORGANS!

Time did not permit of my gleaning any particular book, but only to look through one or two catalogues concerning musical matters, and to get an insight into the system upon which the Libraries work. But I have booked in my memory certain books which I will inevitably scan when I have plenty of time to spare, namely Saturday between lunchtime and 3 o'clock the time of the notorious Recital. (The reference library can be seen on your postcard at the rear of S. Georges Hall).

All appears to be above board at India Buildings, as regards the Union-Castle Mail Steamship Company Limited, and as I have said before, things do appear to be much more settled. The large stocks of stationery brought to these offices of ours from Southampton, portend no early return, and the furnishings of some of our departments display an intention of permanency come to Liverpool to see the Shipping Companies. Of every description from the humble ferry boats to the gigantic liner, their representatives are in Liverpool!

Having said quite enough for one week's epistle, I do at last conclude this letter, after nearly forty minutes typing. 'the time is now 48 minutes past two, and the next item on the programme...' continues the radio announcer.

Hoping this reaches you in the best of health, happiness, and safety, as I am pleased to say, it has for me.

One further word. I forgot to say what I had for lunch today at M. Lyons. hare soup; roast beef, Yorkshire pudding, chipped potatoes, sliced carrots; strawberry ice cream; and coffee (time is now 2.56)

From your affectionate son,

P.S. Thank you for your letter. Parcel enclosing 3 shirts, socks, and collars

* * *

Letter 13

1, Manville Road
New Brighton
Wallasey
Cheshire
28th October, 1940

My Dear Mum,

With reference to your letter in the parcel, enclosing the vest and pants, socks and 'yaples', for which I thank you, there appear to be many kinds of guns here. Some guns are silent, only a swish of the shell tells us of its action; others sound like doors banging, some like a flat wooden board falling to the floor with a sharp crack, and others prefer to boom out their destruction. It is like a new type of organ. Instead of picking out the Diapason tone, the reed tone, the flute, or string tone; one picks out the swish, the clatter, the bang and the boom noises.

I guess my poor father had his ears chewed off (to put it vulgarly) for interrupting you in your letter. Before I had read your excuse – 'Dad keeps reading out bits of the paper...', I myself began to wonder whether it was the noise of the guns going around you and some other malicious disturbance that was the cause for your inaccuracy.

I received your first parcel in my 3rd term on the 22nd inst, 5 days after the date of its letter. Your last parcel, dated the 22nd (the day I received your first) was received by me on Saturday, the 26th inst.

I am pleased to hear that you have been to Millbrook Church and have seen Mrs. Reeve and Mr. James. My friend Mr. Hume is in Southampton this week on a week's leave – lucky him. I am told, that Mr. Hume, has some property to let in Highfield, I should make inquiries if I were you, should you desire to be nearer the centre of town.

I believe I may have forgotten the postcard to Grandma had you not reminded me, so I must hasten to fulfil my duty, although I am afraid it will be late. When I look through my past letters I begin to wonder how I can think of all the things I do to say in them, although while writing this, it is only 3 days since I threw out my last three-page letter, I still seem to be able to think out more and more – it must be the never-ending train of thought!

Having dealt with your last letter fully, I will now proceed with fresh matters. You will, no doubt, remember me mentioning my intended visits to the Ref. Lib. At Liverpool, in my last letter, dated 25th. Today, on Saturday, I carried out my intention. C.A. Rice, having nothing better to do, chose to accompany me. I mentioned the required subject to the Librarian who looked up references to their vast index system. The vast edifice is circular, stacked with several million tomes throughout the entire circumference, from the floor to the domed roof. I took a seat before a table, and presently a boy appeared with three books. The first book, on top, I eagerly grabbed as it was 'The Organ' which fascinated me so much that I had no time left for dealing with the other two. I searched through the noble pages until I found notes dealing with the organ in St. Georges Hall, and found the specifications, which I proceeded to copy out, and enclose herewith a copy, for your information. Later in the afternoon, there was a period of alert and the A.C. had sounded.

I, then proceeded over the road to the S.G. Hall which incidentally is opposite the Ref. Lib. C.A.R chose to follow me to the recital, a copy of the programme I enclose. The recital let a yearning to attend further recitals of equal brilliancy.

On Saturday night, I had my first organ practice of my 3rd term, opening as is now my habit, often such long intervals, with National Anthem. I allowed myself 30 minutes to go over revisionary work, like a pea in the Pacific was that 30 minutes! I want Hours! – Time himself! That night Jerries passed over in waves for three hours during which time our defenses fired back only two shots, hence he was not disposed to send any back. I went to bed at 10:00 p.m. (two hours after the warning had sounded) and did not wake up until morning. I was later told that the A.C. went at 2 o'clock.

Your affectionate son,

Edwin

***

Letter 14

1, Manville Road
New Brighton
Wallasey
Cheshire
1st November, 1940

My Dear Mother,

I, hereby open my fifth letter to you in my 3rd term in Liverpool.

I thank you for your two letters, dated 28th and 30th inst, received on the 30th and 31st inst, respectively. In your first letter, I was rather surprised at your remark that you had not received a letter from me. Actually, I had posted two, one on Sunday and the other, a parcel on Tuesday morning at 9 o'clock. Then I received your letter on the 30th saying you had not heard from me. Of course, I know that possibly by the time I received yours, you had received one of mine. The following is a list of dates concerning our correspondence, which I think will answer your remarks.

Third Term

Mine	**Yours**	
Dated	**Dated**	**Received by Me**
15. 10.40	17.10.40	22.10.40
18.10.40	22.10.40	26.10.40
25.10.40	28.10.40	31.10.40
28.10.40	30.10.40	31.10.40

It is admitted that there is a boat of ours at home, but this does not, of course, herald the return of the mail-fleet. Although there are many inconveniences. For some lengthy period, H.O. have removed from the metropolis, to a port in Scotland.

As a result of this stable attitude, I desire to make provision for my music, to run as a relief to my work. As I have told you there is no instrument where I am now, and furthermore, Jerry hinders me in going to my Musical source, namely, St. James.

Things are very busy at the office. The Vic. Supt. is here for three days to assist the boat for this week. It is some while since we had a big boat here, as most weeks they have gone further north. It became so regular in the last respect that we anticipated a further move in the same direction. But doubts are removed, when we see on our private movements list the imminence of two arrivals for this month.

Victualing has been good (I may venture to state the coveted mark of excellence!) the outstanding 'home' feature is an addition to the supper, a course of soup. I took the liberty of enumerating the quantity of food for me to devour at one particular meal during this week. i.e.

BROTH -Mixed Veg, all possible.	HAM – covered plate
TOMATOES – 2 whole (placed on ham)	BEETROOT – took 3 slices, and HALF SLICE BREAD
BREAD & BUTTER – 5 slices	CREAM CRACKERS – 3
CHEESE (CREAM) – 1. Cheddar	BISCUITS – 5 assorted
COCOA – 1 CUP	SEASONINGS – mustard, pepper & salt

Raids for the past week, have been up to the resent (since my last letter), quite 'peaceful'. We often get a warning at either 6:30 p.m. or 8:15 p.m. or 10:25 p.m. We usually get a period of days when he starts at the same time. Yesterday was an exception (Thursday, 31st of October) there was not one warning through the day or night! But today while having lunch at M. Lyons the alert sounded, but I did not hear it. I saw a swarm of people outside Woolworth's which is opposite the café. I might also tell you the name of the Street – it is Lord Street, very like our High Street, yet physically an exact copy of Commercial Road; having a valley in the middle, and turning to the left as it ascends.

Earlier in the week, the atmosphere was fresh, the water choppy with the addition of wind and rain on Wednesday. On reflecting, there has been a great rise in temperature, for it is as warm today as it was the day I arrived on that fateful August 11. I expressed an anticipation to Mrs. Hart that we should possibly experience colder winters here than the South. Here, she gave a few striking instances. According to the conversations it is summarized as follows. Often the winter sets in earlier here but we do not get the extremes of heat and cold as you do. It is a very rare thing indeed for them to have snow, with the exception of last winter.

Moody is in Bournemouth today, on Friday, if he proceeds according to his plan. Having spent 7 days, he has only 2 more to come. C.A.R. is on the train now having had two hours travelling, for it is, as I type this line, just 2 o'clock. Next week, it is my week. I look forward to this monthly travelling. It is exciting travelling on long train journeys, that never seem to end.

Well, see that your time-pieces, your cats (sorry, I believe one or both is mine!), your bird, and dad, are all in working order when I come, including my faithful bicycle. I must write to Mrs. Reeve to avail myself of her kind offer to let her know when I am coming home. The sole purpose of the invitation is as a conference – to put it in her words. – 'next time you are down come to me on the Saturday to tea and talk things over'. I must also see the stampers, Stuart, Jim, and all the other cronies.

Your affectionate son,

[signature]

P.S.
May I be permitted to make yet another page. To show the extreme freaks of happenings on the 31st of October, and the 1st of November.

As before reported in this letter, there were no raids on the 31st, Thursday morning or evening. Now on the 1st, Friday, we had a raid during the lunchtime. Another one came on at 5 as I was leaving the office with the intention of having another practice. It looked then that I was to be disappointed. Yet by the time the boat had reached Seacombe, the A.C. went. I proceeded forthwith by bus to 'home' to pick up my music, which terminally lives in my attaché case. I was then able to have my second practice of my third term, another 30 minutes! How delicious! I played the following:

1. *The National Anthem*
2. *No. 1 Prelude in No. 1 book – Bach*
3. *3. No.3. "   "   "   "   "*
4. *Wedding March   Mendelssohn*

My technique does not seem to have lost much of its polish despite of the lack of constant practice. Once finished, I proceeded to the Choir Practice, which finished at 7:45p.m. on opening the door sounds of the 'music' floated in – the warning.

Speedily moving with a member of the choir, I was going along Seabank Road, when I heard Jerry. Gun opened up all round, in their usual manner. The luminous effect was a maddened thunderstorm like many flashes of lightening springing from everywhere. The noises were of course more deafening, as I was out in it. We ran at a terrific rate until I reached No. 1, Manville Road, to get away from the shrapnel. Blue flashes denoted bombs dropping in Liverpool. When the firing had ceased my friend of the choir proceeded on his way home. While having one of my glorious suppers, Jerry returned again. Guns opened up – then – wheeee – wheeee – wheee – whoomp! – whoomp! – whoomp! The ground shook as if by an earthquake, the doors and windows rattled, and New Brighton received six more wounds. I sensed the direction in which the bombs had fallen – somewhere round the church. I have not yet had time to see the damage, but Mrs. Hart confirmed my accuracy. Although she mentioned some such buildings as the Tivoli, which I know to be not very far away. The bombs dropped at ¼ to 9.

Another freak. Within three-fourth of an hour, the A.C. sounded. Then the rain pelted down in torrents. It has been raining since noon. Beating against the windows, it ran like rivers down the road. A flash, then a clap, denounced the arrival of a THUNDERSTORM. The lightening flashed, the thunder rolled, and the rain poured for 30 minutes. I went to bed and fell asleep until it was time to get up.

* * *

Letter 15

1, Manville Road
New Brighton
Wallasey
Cheshire
12th November, 1940

Dear Mum,

The train for Basingstoke steamed in at 9:00–10:00 p.m. last night, Monday. It arrived at Basingstoke on time to meet the train for Reading. We arrived at Reading by 11:30 p.m. the Birkenhead train rolled in only 10 minutes late, that was at 1:20 a.m. The majority of the seats were occupied, so I was unable to lay out across the compartment as I had done before. We remained in this train for 8 hours travelling all the while. I had a patchy sleep. Not feeling very awake, I slowly after fumbling, raised one of the blinds of the compartment. On looking out, I saw it was getting light, but there was not much of it. It was a dreary countryside spectacle that met my eyes. It was somewhere in Cheshire, the rain pelted down, and the wind blowing the bare limbs of the trees, which endeavoured to resist the gale. The train arrived at Woodside Station, Birkenhead, at 9:15 a.m., and after a rough crossing by the ferry to Liverpool, I arrived at India Buildings at 9:30 a.m.

This commences my 4th term of musical imprisonment in Liverpool. It is not raining as I type this letter but there is a gale blowing up the river whipping up some furious waves under a cloudy and stormy sky.

Hoping Christopher [budgie] has recovered from his madness and that he is now free from soot [he flew up the chimney], and cleaned and spruce as I saw him when I arrived. I thoroughly enjoyed my 3rd leave spending Saturday afternoon with Mrs. Reeve and then attending services at Millbrook.

Enough for the present, hoping you got home safely and that dad has not released the cats and got into trouble for it!

From your affectionate son,

Edwin

* * *

Letter 16

1, Manville Road
New Brighton
Wallasey
Cheshire
15th November, 1940

My Dear Mum
,

I thank you for your letter, undated, received by me this morning during breakfast. I often recollect the few days I spend on leave each month. The musical items stand out with indelibility, to such an extent, that I can almost smell the Hampshire air.

I am afraid I must severely criticize your sentence regarding the incense and cabbage. You say the smell of the incense outlived the smell of the cabbage. I detect the odours from cabbage being given the sweet and delicious name of 'scent'. If that is the notion of perfume – I fade out! You know how I love cabbage!

Since I have returned there has only been one burst of G-fire, during the ten minutes which it lasted I took the trouble to count the separate 'bursts' – I reached 85. The weather has been freakish, including hail storms with thunder yesterday. The victualing at the digs has been up to the usual standard. The supper of a few nights ago commenced with Fish and Chips. Last night, it commenced with soup.

Two battles were attempted on the Chess board on Thursday. Both were successfully for me, but in the first battle I was in the act of taking the opponents knight when the wretch bit me, and slipped from my finger and thumb, knocking three other pieces from their positions. The opponent flew into a fit and viciously shook the remaining pieces from their positions. We started a new battle, and this time the opponent laid his head on the table 'to get a better perspective of the game' and gradually drew the tablecloth across the table, neither of us noticed the chess board increasingly over-lapping the table, and suddenly the whole issue flew to the floor, hurling 32 pieces to death on the rug under the table. I sank with my comrade into fits of laughter, then collected the unfortunate pieces, and packed the chess for the night.

So far, I have not done any music but I hope to open up by attending the choir practice at St. James Hall with its new member, Mr. C.H. Moody. Tonight, will be his night to enter a new career. Tomorrow, it is doubtful whether I shall attend the organ recital at the s. Georges Hall, if there is one, as Mr. Smith seems only willing to give me a lesson in daylight.

The photos are very good. I have shown them round the office this morning and the verdict is unanimous.

I cannot think of anything else on which to report as there are only the three days to describe. So, I will close now, until I hear from you again next week. Hoping you and dad are still as fit as ever, and as safe, and Felix and Christopher, as you all were when I departed. I am now recovering from the long journey, and am almost living a normal life.

I am, your affectionate son,

P.S.

I regret to state that there was no Organ Recital at the Hall this day so I made full use of the time to hand. I arrived at St. James Church at 2 o'clock and is my custom commenced a recital[ii] to myself with the National Anthem. An organ Lesson followed at 4:30 p.m. and finished at 5:30 p.m.

***

Letter 17

1, Manville Road
New Brighton
Wallasey
Cheshire
22nd November, 1940

My Dear Mum,

In commencing this letter, I trust that there is some form of communication in the post for me, from you, as I have not received any communication since Thursday, the 14th inst. If it be a parcel, the delay is readily understood.

Members of the staff who went home last weekend 14–17th inst, came back with many stories of destructive raids. News reached my ears that one raid was for 13 hours, from the Sunday evening until the Monday morning, all the while high explosives being dropped, including a number of land mines. I also heard that many districts of your town were visited. I trust you are not affected by this. One of the members returning from home, took 23 hours to make the journey, being bombed when he started, and again at many points along his precarious journey.

A vessel of ours has returned to a port in the kingdom, a little knocked-about, mostly by heavy seas. There is another knocked about in another way, but to no great extent. The 550 lb explosive aimed at the latter, pierced the for deck, and continued its career until it reached the smoke room, then it rebounded into a corridor and ricocheted in another direction. Its path was traced by a large hole in the smoke-room and twisted metal along the corridor and cabins. Lord Haw-Haw's statement that the ship was hit was correct for once – but not as much as he had expected – for it did not go off!

Some nights have been a little noisy around here, made mostly by the defenses. The weather condition, I feel, have helped as it has been inclined to be rough. The heavy swell on the River tells of great tempests out at sea. With one vessel here it makes us busy, allowing little time to do anything else. We are expecting another big one in here early next week, which will make two mail vessels, and two cargo vessels of ours here. Sometimes we have none in at all, now we have four at once.

My friend Mr. C. Moody, attended the choir practice and immediately became enthusiastic in the work. The practice lasted about an hour, during which two familiar anthems were rehearsed – 'The Radiant Morn' (Woodward) and 'Sunset and Evening Star' (Parry). I recalled the time when I sang this in Winchester Cathedral. Mr. Moody's first Sunday in St. James Parish Church enveloped him with musical interest. A practice is held after each service and the latter Mr. Smith promised an attempt to do a number of choruses, etc. from 'The Messiah' in preparation for Christmas. The next day, Mr. Moody visited a local library and produced on his return the copy of Handel's famous work. When I will come home next time, I must bring a copy back with me, but until then I will borrow his borrowed copy, especially this weekend as he has gone on leave today. Last Sunday evening, the anthem was 'Like a Father' by Hatton.

I have been very busy with correspondence both at the office and at 'home'. At home I have written to Mr. Stephenson and Mr. L H Torr. On Tuesday, I received a letter from Nordic, and replied to it the next day. Later, I received a letter from Auntie Ruth which I have yet to reply to.

I must now close my third letter of the fourth term in Liverpool, hoping that you are unscathed by the devastation that I hear is going on 'in the South of England'. I shall be pleased to receive a parcel containing a shirt and collar, or otherwise I will have to supply myself with an additional one from a shop in Liverpool!

I am, Your affectionate son,

Edwin.

***

Letter 18(a)

1, Manville Road
New Brighton
Wallasey
Cheshire
27th November, 1940

My Dear Mum,

I was so pleased to receive the parcel from you enclosing shirts, collars, vest and pants, chocolate and nipits [liquorice sweets], which were awaiting me in the hall at the above address, on Monday night, the 25th inst.

My fingers have not touched the organ since the practice last detailed to you. Last Saturday, there were no Organ Recital at the Hall, and it would have been no different for me even if there had been, for I had to work through Saturday lunchtime, and on to 5:30 p.m. in the evening. I have remained within doors every evening except Friday, when I go to St. James Hall for the choir practice. Many of our Dept. worked all day Sunday.

For the first time, this month I attended a Cathedral Service. On arriving in the Cathedral, I heard the organ, releasing many tones. The part of the Cathedral formerly used by the choir and congregation is still out of use, and the part behind is still being used. The choir is placed nearest the east, next to the altar which is against the screen, and the congregation have been placed opposite each other under the tower. The choir entered from the back, at the North-west corner, and crossed to the centre and turned and proceeded to their stalls in the East, passing through the middle of the congregation. The organ ceased on giving the chord for the preparation, the Introit, one which we knew well 'Lord for thy tender mercies' – Farrant. This was sung unaccompanied, giving Goss-Custard time to descend from the Organ and enter through a small door in the northside of the tower. I was surprised later, after the Introit, to hear the piano, and on stretching my neck I saw Mr. Goss-Custard at a Grand, to the left side of the altar. There was also another anthem, 'The Lord hath been mindful of us' – Wesley. The choir was dismissed, and Goss-Custard had returned to the organ loft and gave us a great concluding voluntary, filling the entire Cathedral with music. Thus, ended a charming service.

Yesterday, we still have been without a warning for 3 days, and we had none again last night. At the office, things are still very busy, getting one boat out, after only 4 days in port, and receiving another one. We received a message from General Dept., in letter form, regarding leave. I will go to see if I can find it, or otherwise I will have to memorize its contents. ...here it is... This is the part that governs me.

'SALARIED STAFF' (2) married men, whose wives are in Liverpool, or single men, who wish to go to their homes, may...each six or eight weeks...' at the moment, I do not know when I shall be coming down next. It may be six weeks from now, or still from last time. I would have been coming on the 6th December, but may now have to wait until 20th prox, or even until next year.

I am enclosing a fortnight's washing, and trust it will get through the post already and returned safely.

Trusting, Dad and You, Felix and Christopher, and Peter are all well, safe, and unscathed. Although, I hear that you have had several severe raids since you wrote me on the 19th inst. One German report in a newspaper stated that 250 planes took part last weekend, dropping 300 of High Explosives, and 1000's of incendiary bombs on Southampton.

I am, Your affectionate Son,

Edwin.

***

Letter 19

1, Manville Road
New Brighton
Wallasey
Cheshire
29th November, 1940

My Dear Mum,

I do not expect that my last communication, in the form of a parcel, posted on the 28th, will have reached its destination by the time you receive this letter. That parcel contains 2 shirts, 4 collars, socks, vest, and pants.

On the 28th, Thursday night, I obtain the key to the under hall of St. James Parish Hall, where I had my first piano practice since I left Southampton, the second practice at a horizontal instrument since my return on the 12th. I remained there from 6:00 p.m. until 7:20 p.m. I am hoping to have another practice on the organ on Saturday.

Before I reached the end of the road, when returning from practice, the air raid warning sounded for the first time in 5 days. I reached home before there was gun-fire but from 8:00 p.m. that night until 4:00 a.m., the next morning the guns pounded away at the various targets. About midnight, a number of flares were dropped. High explosives were dropped in some districts, but many miles from New Brighton. Although it is claimed that one bomb was dropped by The Tower by the promenade. Damage was done in the City, but as I am not particularly interested, I have not bothered to go out to Childwall to see it (about 4 miles from the office).

When returning the parcel of laundry, I should be glad if you would send 'The Messiah', as there are no prospects of me coming down to fetch it. I would advise registered post.

As I am pushed for time, and also subjects to relate, I close this letter, hoping that you are in the best of health, and in perfect safety.

I am, Your affectionate son,

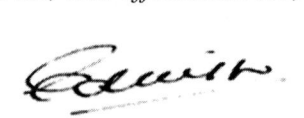

P.S. 30.11.40

I have had very little time to compile a letter this Friday – as I wrote most of the week's news in the letters in the parcel, dated with, and posted at 9:15 a.m. on the morning of the 28th. There being only three days left on which to report with further ref. to the air raid here, there were 234 casualties.

I thank you for your letter on the 25th, posted 27th, and received by me on the 28th while at breakfast. You certainly have been having a bad time of it in the South, and I was rather anxious when I heard of these mass attacks being nightly. I have heard about some of your damage. St. Luke's Church, [Southampton] having the back sliced off it – I expect Mr. Duncombe at the opposite corner had a rough time – the Picture House, Classic and Forum Cinemas have been affected, the story with windowless Commercial Rd; Bitterne seems to have had several surprise attacks. It must be fearful to be in the thick of it, but it could not be so bad to any particular person unless it is as near a miss as the Millbrook packet was to us.

I enclose a copy of the news we received from our 'G' Dept. concerning leave. I anticipate spending Xmas Day in the work house i.e. India Buildings, as there is a prospect of much activity around that date. For a practical joke on Xmas Eve, I suggest suspending by a dainty coloured cord from our own bedroom door, one of my time-worn socks, to tempt something from the landlady. The present, no doubt, would be additional laws and regulations regarding the misplacing of personal apparel.

I have enlisted about 30 odd persons to be honoured with Xmas cards this year, on which I hope to commence work this week.

Sunday morning, I once more sang it S. James Choir, the frosty air suiting my voice admirably. It seems such a long time. I was able to 'let rip', with my soprano voice at the ages of 12, 13 & 14. For a period of four years that voice remained dormant, now it is waking to its new tenor life.

***

Letter 20

1, Manville Road
New Brighton
Wallasey
Cheshire
6th December, 1940

My Dear Mum,

I am so glad to receive your communication this morning. The card and the parcel arrived together. It relieved my anxiety, to hear that you are all safe – but what a near one! What a terrible fright it must have given you when the glass and the ceiling gave way. Were you in your table-shelter at the time? Another house of ours wrecked! While 'Norwood' moves to another stage of devastation?

Your letter was very interesting, and by contents it appears that my home town has disappeared from the map. Millbrook must have had a warm time of it as well, with G. Morton's being hit and by Colebrook's old house. I fear I shall not see the Soton I left behind on the 12th of November – Oh dear! I shall not know the place.

We have had a quiet time for a long time with the exception of that one almost ineffective Blitz of a week ago. It is just too quiet for my liking, it portends of nothing heathy in the future.

On Wednesday evening, a rash broke out on my dial that put the wind up me. But on Thursday morning, the scarlet, irritating patches were on my legs and arms! So, I visited a doctor, before I died of fright. It was a nettle rash, and I am now taking two things (a) the medicine, and (b) a couple of days off! Greedy me! I want one thing more a medical card! I believe you have the latter. But, in the meantime I am trying to get a temporary card here. The rash is less already.

Now I wrote a letter to Vera only last week and you have taken the trouble to write another one! Gosh! She'll be busy answering that packet.

The Anthem last Sunday at New Brighton went well. There were 10 Tenors and Basses, later joined by 13 boys. After the service, another practice was held in the Hall, when we had a shot at Handel's Messiah'.

Early in the week we had frost, but now we have a gale with rain. I had hot lunches on Sunday, Monday, Tuesday, Thursday and Friday: all this week.

I enclose the copy of the memo that I said I was enclosing in my last letter. I was in a hurry to catch the post, last Sunday.

Yesterday, I sat in reading all about 'Another man's wife'. This morning I have spent scrawling this letter. Early in the week, I sent a reply to Auntie Ruth's. Hoping you will have no more raids and that you will not suffer from your experiences of the last weekend, and other preliminary attacks of the last fortnight.

From your affectionate son,

Edwin

***

Letter 21

1, Manville Road
New Brighton
Wallasey Cheshire
13th December, 1940

My Dear Mum,

I returned to the Office on Monday morning, the 9th inst. And reported for duty, after a short break of three days sickness. The rash bore the name of 'Urhcaria', or in more common language, Nettle Rash. On the third day, last Saturday, all traces of it disappeared, although I still feel a little indisposed. But, nevertheless, I spent the morning and afternoon of that day on the Organ in St. James Church, playing my old pieces.

It has been a very busy time for the past three weeks, at the office, as there are so many 'tubs' in at once, and all require victualing. There is, therefore, little time left in which to type this letter, but seizing a brief spell, I may be permitted to gather together, a few scattered thoughts, and set them on paper.

An item of paramount importance to me this week, has been the question of laundry. I find the stock of shirts etc. inadequate for the periods which I, now have to stay in Liverpool. To bring this matter up to the standard, and quantity of others of the Staff, I require to know the size of shirts and socks, and all sizes of my apparel, to enable me to obtain the necessarily larger stocks. In your letter of reply, I shall look for a list stating these sizes. This will enable us to go a longer time before having to bother about posting.

Last Sunday, the 8th inst., two Anthems were sung. In the morning we sang 'Like a Father' (Hatton); in the evening we rendered 'I did call' (Moir). The Te Deum in the morning service was sung to a setting of H. Smart in F.

On Wednesday night, the guns spoke for the first time for 14 days, during which period many days passed without warning. For at least half of that time, there were not even night warnings.

I have received one other letter during the week, from Auntie Ruth, which is still awaiting a reply. Hoping you are now getting straight again, although I do not know the extent of your total confusion, perhaps one day, I shall be able to see what remains of it, myself. Also hoping that you are enjoying the best of health, with my father, and trusting you are comparatively safe.

From your affectionate son,

P.S.

Since typing your letter, I have had a great pleasure in receiving your parcel, and your letter dated 12th.

These most interesting pieces of correspondence provide me with ample food for digestion, regarding the effect of the Blitz and the many results occasioned. I thank you for sending the linen, newspapers, Messiah, and a medical card enclosed with the letter.

Last night, another peaceful night I spent endorsing Xmas cards etc. which I must say took up the whole evening. On the afternoon, I had another practice of St. James.

***

Letter 22

> 1, Manville Road
> New Brighton
> Wallasey
> Cheshire
> 20th December, 1940

My Dear Mum,

I certainly have been indulging a little in work for the past month or so, at the office and at home. The latter has consisted of the matter of Christmas cards etc. Let me give, below, a list of those who have been favoured with such communications.

Posted 17th Mum and Dad (reg. Parcel) 17 Northlands Close, Totton, Hants

"17th		Aunt Ena and Tizzard.	'Francsland', Water Lane, Totton
"	"	17th Grandma	-do- -do- -do-
"	"	17th Aunt Marion and Bert	35, Mansion Road, Freemantle, S'ton
"	"	18th Aunt Ruth (& letter)	20, Clifton Road, Parkstone, Dorset
"	"	18th Aunt Beat & Cousins	12, Dudley Street, Nr Russell Park, Beds
"	"	18th Stuart Durley (&let)	24, Meadowmead Ave, Millbrook, Soton
"	"	18th L.H.Torr	5, Church St., Hay, Herefordshire
"	"	18th R.W.Stephenson	Docklands set. No 5, York Blds, Soton

Posted 18th Nordic Shepherd (With calendar) 23, Hiltingbury Ave., Chandlers

"	"	17th Vera Hollett (with calendar)	77, Rochester Ave., Earlsdon, Cov
"	"	17th Mrs. A D Reeve (with Calendar)	3, Archers Road, Soton
"	"	17th P. Pickering	124, Upper Shirley Avenue, Soton
"	"	18th Rev. Beaumont James	The Rectory, Regents Park Road, Soton
"	"	18th Rev. Molyneux	S. James' Vicarage, Dockland, Soton
"	"	17th Jim Hann	72 Imperial Avenue, Soton
"	"	17th Mr. & Mrs. X Rex Hollett	77 Rochester Rd., Earlsdon, Cov
"	"	Mr. A Guest Smith	c/o S.James Vicarage, New Brighton, Ches.

I expect that you have received your parcel by now, which as you see is at the top of the list, that it was registered. The postman will require your signature for the parcel, so should you be out when he comes, he would take it back to the Post Office and bring it again, and again until he catches you in!

There have been few warnings since I last wrote, and the weather has been good for the time of year. The question of my leave is still in abeyance, owing to the amount of work going. I cannot go into details as to why we are so busy, but that will be something to talk about when I do come.

Wishing you the full complements of the Seasons,

I am, your affectionate Son,

P.S.
22.12.40

Jerry has provided me with a P.S. this week. On Friday, the warning provided at 6:40 p.m. just on my arrival at home from the office. Within 10 minutes, Jerry's engines were heard in the distance. This was the first time I had heard them for about 3 weeks, when they used to pass over to the Midlands. Therefore, on hearing his approach, we guessed the same again, but before he was overhead, the screech of bombs told of destruction on our side of the river. During previous raids, New B. escaped with only six or seven bombs, so thinking that all was over, I drew out from my improvised shelter. Another Jerry was following the firsts and he, too dropped some nearer this time. Before the warning had been on 20 minutes, things became alarming, as all the Jerries concentrated on New Brighton; large numbers of incendiaries fell, 1-2 doors away. Thus, the raid became more infuriated as more Jerries poured in. The sky was illuminated by the many fires, and made more fearful, by the flashes from bombs and guns. Moody and I took our positions under the stairs with Mrs. Hart and Co., as it seemed a little safer there. With but after a few breaks, the raid continued until 2:00 a.m. in the morning but A.C. did not sound until 4:00 a.m. but time bombs went off during Saturday. I went to bed at 3:00 a.m.

P.T.O. for what I saw when I left to go to work Saturday morning[iii]*.*

***

Letter 23

Sandrock Hotel
Seabank Road New
Brighton
Cheshire
24th December, 1940

My Dear Mum,

In pursuance of my last communication, posted Sunday, the 22nd inst., I must have given you the impression of the happenings of the last few days. The two blitzern were directed mainly against Liverpool, and the district in which I resided. On the Friday evening, the first Blitz things were pretty hot around us, and over the other side a bomb demolished 4 dwellings between the Adelphi Hotel, and Lime Street Station, and a bomb landed in the middle of the road at the end of St. George's Hall. Remains of the incendiaries were to be seen everywhere, including around India Buildings.

The following day, on Saturday, we had two warnings in the morning as a preliminary to the evening's entertainment. In the afternoon, I attended the last organ recital at the St. George's Hall, the building being gutted that night. At 7:10 p.m., the warning went and within a few minutes, Jerry flew low over New Brighton, dropping H. E's at the rate of one a minute (dropped in eights every two minutes). This continued for 9 solid hours, without more than two minutes respite at midnight. As you can see by the diagram, I drew for you in my last letter, that many fell quite close to No. 1, Manville Road. The fear fullest crash of the lot was the one that fell in Seabank Road. The house was filled with dense clouds of dust, soot, sand, and smoke. Fortunately, we were under the stairs at the time, for all the back windows and frames flew in. the houses near that bomb lost their roofs. Well at 4:00 a.m., I retired to bed, and did not succeed to wake before 11:30 a.m., the next day.

Sunday was a day of frost. Managing to extricate myself from the bed, I dressed and washed in the windowless bathroom, the icy wind blowing through, giving me the shivers. The old sound of tinkling glass and sweeping filled the deserted streets. The first sight, we saw (moody and I) on commencing the walk around, was to see the damage done by the bomb close at hand.

We went down magazine Lane and found Vale Drive roped off and so continued down to the promenade. On the way, we found many roofless, windowless houses. The cause of much of this was a 1000 lb. bomb that landed in Vale Park. The crater was enormous, about 26 feet deep and twice as wide as the ordinary road; this was not much more than 100 yards from No 1. Manville.

I had made up my mind to go to M. Lyons, in Lord Street, but on the way, I passed many smouldering ruins, and turning down into Lord Street, found to great fires opposite M. Lyons, and that part of the street roped off. Trying to get back to the office through short cuts, I found every way balked by fires, smoke, water, and rubble. I had to content myself with a cold lunch.

Now I come to the point in this muddled history, when I have to explain my new address. On arriving, I found three cards for me and had just picked them up when Mrs. Hart appeared carrying in coals to make a fire for us in the front room. She opened up by saying to Moody and myself, "I am sorry, but I am afraid you boys will have to find somewhere else to stay as we have to get out ourselves". Mrs. Hart, herself is removing today to another part of the country, as are many other people around here. So, we straightway pushed off in search of digs; all houses being windowless etc. for miles around it was a difficult problem to arrange accommodation in so short a time. A member of the Choir also an evacuee from the South, happened to be staying at the Sandrock Hotel, which has had its ball room bombed, but nevertheless it has one vacant room, which will accommodate Moddy and myself for a week or so.

By Tuesday morning, however, we had packed our belongings given over the key, and had my credit of boarding fee returned to me, and had taken our cases to the Sandrock Hotel, which will now be our home for a time.

It is going to be very difficult with the mail problem, for I wrote the No. 1 address on all my Xmas cards, letters etc., so that should they be undelivered, they could be returned to me, and now everybody will be sending their stuff to a vacated address. The only clear solution is that I must call there periodically to collect same.

I managed to get a lunch at Maison Lyons, as the barrier was moved up one place to enable the café to be used. I had Oxtail soup, veal, chipped potatoes, and carrots, Strawberry ice cream and coffee. We were able to look through the window (which incidentally was entire) and alight our eyes upon the divested, gutted four-storey buildings, of Phillips, Jays and Broad bears, covering an area of several 100 square yards. Smoke was still visible from these ruins.

Trusting that you had a quiet week end, judging you should have had, considering so many thousands of planes were engaged upon us, and trusting you have had some sort of a Christmas, although I am afraid Christmas for me, and others will only be a day of recovery from the tempestuous treatment which we have recently undergone. I must close this epistle, which closes down the old address and opens up the new one.

From your affectionate son,

***

Letter 24

Sandrock Hotel
Seabank Road
New Brighton
Cheshire
27th December, 1940

My Dear Mum,

I have made arrangements to collect 'the daily mail' from the house next door, in Manville Road, and yesterday I was pleased to find the little parcel from you. I unpacked it in the Lounge of the Hotel, on Boxing Day evening, and was pleased to find that it contained a watch. It keeps perfect time, and I place it on the chair beside the bed at night, which saves me getting up in the morning to find time, often finding myself waking at 3, or 5 o'clock in the morning. I also thank you for card and letter, dated 19th.

Well I had quite a good Christmas, considering the upsets of the previous week end, and transport difficulties during the week. On the Festive Day, I was requested by a member of the choir, who is staying in the same Hotel, to give them a piano recital. Thinking of nothing more suitable, I commenced with a few bits from the 'Messiah'. And then concluded with a number of Bach Preludes and Fugues. The number present were 40 or over. The later recital included a number of sight-read minuets of Beethoven. I was so glad to be in a position to sit at a piano, that I did not heed the presence of the others, who appeared to appreciate my efforts.

The Victualling over Christmas included a fine six-course lunch, the third course including Turkey. I am now able to have meals at the Café in Liverpool again, today, Friday, I had Bagartian Soup, Roast Mutton, chipped Potatoes, mashed swedes; Ice cream and coffee.

I forgot to mention that there was one disturbance on Christmas Day – the fact that a mine went off at 5:45 p.m. on the prom. It made the SAND – ROCK, and blew in their blackouts. It took half an hour replacing the same. We have been subject to these explosions ever since Friday night. We had no warnings on Christmas Eve, or Boxing Day, or even Christmas Day. But we have had a short one this afternoon.

Hoping you had a quieter Christmas, and that you will not get any more blitzern.

From your affectionate Son,

P.S. I can at last see myself having a weekend leave next Friday, 3rd January – should arrive mid-day this time.

***

Letter 25a

Victualing Dept.,
Union-Castle Mail S.S. Co. Ltd.,
India Buildings, Room 320
Water Street
Liverpool 2
1st January, 1941

Dear Mum,

Owing to the illness of a number of members in our Department, and other departments of this Company, my services are required to a great extent. Hence, I regret to inform you that my leave due on the 3rd January is cancelled.

It is again indefinite when I shall be allowed some time off for weekend leave. Up to the present, I am quite well, and enjoying my meals at the Hotel. The Sandrock Hotel is, however, closing down within a week or so, owing to damage. In the meantime, I will endeavour to secure digs in another district, as there are few dwellings habitable in N. Brighton.

From your affectionate son,

[signature]

***

Letter 25b

Victualing Dept.,
Union-Castle Mail S.S. Co. Ltd.,
India Buildings, Room 320
Water Street Liverpool 2
3rd January, 1941

My Dear Mum and Dad,

For the first time, for over a fortnight, I feel composed to write a letter. Circumstances during the past two weeks have been in a very unsettled atmosphere, giving me a strong sense of uncertainty.

It was a great disappointment, for all of us, that I was unable to come home this weekend, but I am glad to hear from Auntie Ruth, that she could not come home this week either; so, we may meet at a later date. Everyone around me are blowing their noses, sneezing, and coughing, and those that are not are away. It is those that are away that are the cause of the cancellation of my leave this week, as explained in my brief note earlier in the week. I am, nevertheless still hoping that it may not be long before I shall be able, once again, to see Hampshire, and to satisfy my curiosity as to the raids-damage. It seems a long time ago now, that I read the papers with their startling headlines concerning your troubles, and I think reason for the time going slower is because we have had a taste of it ourselves.

I have now been at the Hotel 10 days, which I must describe as being windowless (actual number of windows broken is 73) it is a 3-storey building, with about 30 bedrooms. The lounge forms my place in the evenings (when I am not sat in the Dining Room) – or in bed in room 25. The Manageress is a dear old lady, rather stout, but in other respects so like mum, and I certainly feel more at home here, that I did in Manville Road. The fee is 30/- a week, and it is worth it! We have a greater quantity and quality of food than we had at the former address. You remember my remarks regarding the food at that former address, but this surpasses all. For instance, instead of occasional Corn Flakes for Breakfast (1st course), we have porridge every morning. This is followed by <u>Eggs</u> and Bacon or <u>Fish</u>, or something else tasty. Again, the Supper, held at an earlier time than the other place, is <u>HOT</u>, not cold. This commences with soup, then it may be Veal and Veg., or Chicken and veg., or Beef, or Pork, and so on. The Sweet, has composed, Fig Pudding, another night it was Ginger Pudding, another night I remember Apple tart, and another time Christmas Pudding, and so on. The Dinner is concluded with tea. Bread and water are served at Lunch, dinner, and supper.

Now the weekend board is entirely full. All day Saturday – Breakfast, Lunch (Which at the old address was not included), tea and light supper. Sunday, of course, is full board.

To proceed to the bedroom (a very drafty process by the way) is the next item for my discussion. It is funnily shaped as it is in part of the gabled roof; the two extreme ends of the ceiling slope toward the middle, supported by 10" beams. The wash basin, in shape and design very like ours at Totton. The beds are very comfortable, covered with several thick fluffy blankets, and quilt and when I get inside that, I don't budge from it on the slightest pretext. Now you know, how I have been living for the past 10 days – in respect to board.

Music – I have not done much practice in spite of the fact that there is a piano, more or less to save paining the ears of others using the lounge. But there is ONE bright spark in the midst of all, the dismal turmoil!! In extreme contrast to things of the past few months, I had a little appointment! As I told you when in one of my recent letters, I went to St. Columba Church, at Egremont. St. James is closer but the Rector or Vicar or Father (it is high Church) invited Moddy and I to a little tea party. He, Rev. Berkeley, said would like Moody as a server and I as a musician. So, on the following Sunday, I sang in that Choir, in the morning. After that service, he asked me to a great thing – to play the Organ, for the Children's service in the afternoon!

After looking at the organ up a very narrow staircase, I sped back to the Hotel, and enjoyed the glorious Lunch, then sped back again, stumbling into a bomb-crater along the prom, in my haste and excitement, and eventually rolled up 45 minutes before the service, to experiment on the organ, and gain confidence over its mechanism.

The service went well and the clergyman was very pleased with the rendering, especially No. 1 Prelude in C – Bach. The service lasted an hour. (3 o'clock to 4 o'clock). I remained for the Evensong which commenced at 4:15 p.m. and sang in the choir.

I was hoping to come home and tell you all this, but you see, circumstances will not permit.

The next item for discussion is the matter of Air Raids. For the past fortnight, these have been in the form of warnings and all clears, hoping that these will be all we shall ever get. For the first two or three days, after the blitz, Jerries came over in the day to wee what they had done. Last Tuesday night, I believed it was, Jerries, or rather planes, for there is some argument as the nationality, were passing overhead and around us for about an hour, 9:30 p.m. to 10:30 p.m., and the A.C. went. I awoke at 5:15 a.m., the following morning to hear a warning going, but soon fell into a sleep, and started dreaming of a Jerry crashing and the pilot chasing me into the Church where I immediately made for the Organ and jumped down one of the 16' pipes! I could hear the engine ticking over as I thought, and then awoke, and looked at the watch, it was 6:10 a.m. and there was a Jerry going over, and what was more I could hear the bomb release click over and down came six bombs on New Brighton, the first rocked the Hotel like a jelly, and the others fell further away. I flew out of bed, (much to my reluctance) and put on the light and got back again, and listened to the plane's engines dying away in the distance. No more Jerries came over so I curled up again, and 35 minutes afterwards the A.C. went.

The reason for the Hotel closing is owing to damage caused before I occupied that territory, and also that there is a time-bomb under it! So, I'm told.

The weather on the 31st December 1940, suddenly changed to frost, and at 10:00 p.m., there were a few flurries of snow, covering roof-tops like a thick frost. The cold has intensified although the snow has stopped, at present.

I wish my father many Happy Returns of the day, although I am afraid I am little late in sending my compliments but if I had come home this weekend, I would have done it personally.

I should use my office address for the present, as I am uncertain as to where I shall be next.

Your affectionate Son,

***

Letter 26

Sandrock Hotel
Rowson Street,
New Brighton.
Cheshire, 14th January, 1941

My Dear Mum,

I arrived here at Birkenhead at the awkward hour of 2:10 a.m. this morning. The rail journey was good, but I did not appreciate walking six or seven miles from Birkenhead to New Brighton, with the great possibility of warnings, and bad weather conditions. For three hours, I was picking my way through Docks and Quays, directed by numerous policemen, towards Seacombe. Had a warning sounded during my first four miles of the walk, I would have been in an unfavourable position of being surrounded by military objectives, such as mills, refineries, warehouses, gasworks etc. having found my way through the maze of industrial buildings, ship yards etc., I came to Seacombe. I did not meet a single person during the latter course. On reaching the Hotel, I found this locked, so proceeded to the underground shelters under Tower Building. Here were many persons, snorting, and panting, like steam engines, possibly dreaming of the days that they had homes to live in, but now perhaps, more futuristic. My total amount of sleep was one hour, and the additional unnecessary walking was an appendix to my misery. In the future, as in other times, I must travel to Birkenhead leaving at 8:48 a.m., and catching the 1:10 p.m. from Reading in which I can at least have 6 or even 7 hours sleep. And in addition to that there is no walking to do when I arrive, as the office in Liverpool (my correct terminal in town the Railway Ticket includes the fare from Birkenhead by Ferry to Liverpool Landing Stage). Another advantage is that I could get a cup of tea on arrival, as refreshment places are open then.

I made an attempt to enter the Hotel at 7.20 a.m., and after much commotion I attracted a maid to the door. I found Moody in the act of rising from his slumbers, and also found your parcel, which he said arrived on Friday. He went on to say, that he has seen another place in New Brighton, not far from the Hotel; in which the furniture is mainly of Dutch origin, including a Dutch Piano, which he has already reserved for use by me, in case of my consent to the offer. He is fascinated by the place, and says there is a more homely atmosphere than that of our initial lodging. I am to affect an interview on Friday or Saturday.

This letter commences my 5th term of imprisonment in Liverpool, it leaves me as the snow falls on the wet streets.

From your affectionate Son,

Edwin.

* * *

Letter 27

Victualing Dept.,
Union-Castle Mail S.S. Co. Ltd.,
India Buildings, Room 320
Water Street Liverpool 2
17th January, 1941

My Dear Mum,

I have received your letter of the 14th inst., at the Office this morning, enclosing the Collar, which I have been looking for since my return. I find there is another Collar which I should have brought home with my other soiled linen, but this I will send in my next parcel.

It was scarcely believable that the last time last week I was in Southampton, at Totton, in Grandma's awaiting for you to return from Town. It seems to me now, 8 days afterwards, it was as yesterday. I have recollections of visiting the Shelter nightly, shrinking myself in the smallest capacity to obliviate the shivers. On the Saturday I clearly remember cycling to Mr. Darke, the Hairdresser, for the purposes of removing a little surplus hair from the top of my head and also to have same disturbed, ruffled – it has not yet settled.

Sunday again stands out in my memory. The sight of the old church, and grounds, which have seen my growth from childhood, now have occasional glimpses of its veteran choirboy. That memorable Holy Eucharist service, read instead of sung, spoke of the quietude, and desolation of the surrounding parishes. All was still, the life of past years, the lives of many of its worshippers had passed out. The afternoon service, was brightened for a few brief minuets during the singing of the two hymns and the concluding voluntary 'The Chorus of Shepherds' – Lemmens.

I have interviewed the landlady at No. 3, Ormiston Road, New Brighton, with the object of obtaining suitable accommodation. It is, by appearance, with suitable. The Bedroom is in the front, with the Drawing Room & Dining Room combined on the ground floor. There is a Piano in the latter quarter.

I had my first outside Lunch at Maison Lyons, on Friday. Commencing with Paysanne Soup, that course was followed by scrambled egg on toast, with chips, and Swedes. The sweet was Vanilla Ice, concluded by Coffee. At the Hotel, I have already three good dinners on consecutive days. One dinner 2nd course was rabbit, and this was very tender. Another day I had stew.

It has been quite pleasant not having the continual wailing of Air Raid Sirens. I have heard only one since my return, and that was on Thursday night before I went to sleep. I was told that 2 others followed during the same night. Judging from the weather conditions, I do not expect many visits by Jerry for some days.

In my spare time, I am revisiting French, so any time now you may get part of my letters written in that language. In case you have any difficulty over the words, I will direct you now to my book-case, in which you will find a good dictionary. It is, I believe, about midway on the 2nd shelf. Perhaps, later you could send up, the 2nd part of the French book I have with me. It is enclosed in a brown paper cover. Both books are from my days at Clarks College. Que pensez-vous de celaf!

So here endeth the second letter of the fifth term.

From your affectionate son,

***

Letter 28

3, Ormiston Road,
New Brighton,
Cheshire
24th January, 1941

My Dear Mum,

I received your letter in the Parcel, this morning at 11:00 a.m., enclosing the laundry, and other things. The shirt was visible before I attempted to remove the paper, but it does not appear to have gained much dirt.

The principle item of interest this week is my removal from the Sandrock Hotel. On the evening of Tuesday, the 21st inst, I collected my bags and walked over the snow to No. 3 Ormiston Road, New Brighton, with my friend Mr. C. H. Moody. We sat down to High Tea, straight away. There were four sardines on toast, followed by two pieces of white bread and butter, two pieces of brown, and two of homemade bread. Plum jam was available. Two scones and a mince-pie formed the cake. 'buckets' of tea were extracted from the teapot.

In this dining room, cum sitting room, as I before mentioned, there is a piano, a Mignon, made in Rotterdam. The side-board is on the adjacent wall, and a couch opposite in the window bay. The fireplace is on the large size, with two coal boxes on either side, with leather covered lids. On the edge of the hearth rug, is an arm-chair on either side of the fire, two positions in which Chas. and myself are to be found in the evening.

The bedroom is directly over the Dining room, furnished with a double-single French bed. Actually, two separate beds joined together by a blanket over both. Over the beds is cord which works a double action switch for the light suspended from the ceiling, over the bed. This is a great facility. There is another light over the dressing table in the window bay.

The two breakfasts, Wednesday and Thursday, have been porridge, eggs, and bacon, plenty of bread and butter, and jam and toast; and this morning Friday, porridge, kippers and the same bread and butter items. Last night, we had dinner in place of High Tea. We understand, that the night meal will be varied according to commodities available. Hence, we had 'Toad in the hole' – two beautiful pork sausages placed on a 5" square of

Yorkshire Pudding, a mountain of potato mash and a pile of carrots. The second course was composed of a number of green plums with cream. Later in the evening, we were surprised by a further snack of tea and biscuits. The previous night, we had a 'bucket of cocoa' and biscuits before retiring.

The atmosphere and victualing of this establishment places our former (initial) residence, far into the shade. It is truly homely. I have the copy of my last letter to you but believe I mentioned the fact that Mrs. Willis, and her husband, and two boys have lived in Holland for a number of years. Anyways, the result is interesting as much of the furniture is of Dutch origin. For instance my serviette ring is inscribed with the words; 'Ter herinnering ann Dik'. If you care to play with the second word you will soon make up our name – HEriNNerING.

My last two evenings have seen me for an hour each at the delightful piano. The tone is soft and sweet, and quite resonant.

My other activity is at Language. I have done two exercises this week and have thoroughly thrashed out the points contained therein. I am hoping you will send later, the second part of my college

French book, to enable me to pursue those points further, as I have found that the second book deals more fully with each part of the Grammar contained in the first book, and as well as pushing further ahead into more advanced staff. I send you your air raid shelter puzzle for this week, on the attached slip.

I am very pleased to say that since my return on the 14th, up until now, on the 24th, a period of 10 days, there has been only one air raid warning. The longest period of enemy inactivity, 8 days! How we all wish it could continue. I feel sure you would not believe your ears in Southampton should not hear a siren for a similar period.

From your affectionate son,

Edwin.

***

Letter 29

>Victualing Dept.,
>Union-Castle Mail S.S. Co. Ltd.,
>India Buildings, Room 320
>Water Street
>Liverpool 2
>26th January, 1941

My Dear Mum,

I hereby open my first weekend letter since my return to the North. I imagine that you are still having a flood of raids at the rate at which there were demonstrated to me during my weekend leave. Since my return here, I am very pleased to say that there has only been one siren and that was on Friday afternoon. However, on the Saturday I was home they had two raids here during the Saturday.

This weekend Monsieur Moody has gone home on leave, after spending 7 weeks here. I bid him farewell on the steps of India Buildings.

Every evening, I have spent at least an hour at the piano. I have drawn up a rough plan of practice for the forthcoming month. The plan includes much revisionary work.

You will be pleased to learn the landlady came to me one evening requesting my assistance upon her younger son, Rodney, who has once before taken up music lessons but since the air raids (which started September 1940) he has not paid many visits to the piano. He has taken up lessons again with another tutor this week, and immediately many difficulties presented themselves. He came to me, and I think I may say that he has had many of those difficulties removed. He has a first piano lesson book by Carrol. It is similar to the one I had, by Gladys Cumberland.

I find that I have another shirt here, it is one you sent in your parcel, dated the 6.2.41.

So far, I have done little French, but hope to resume strenuous activities in this direction as from Sunday. I am enclosing my usual French News to you.

The question of correspondence is receiving my faithful attention and I will give you an idea of the work before me. I have settled to reply to the letter Auntie Ruth sent the week before my leave. I have received another letter from my friend Mr. Sims. He has been hastily removed from Liverpool to Glasgow. This makes another for me to reply to. I am also thinking of writing to Stuart this week, as I did not see him last time I was home, and there will be no prospects of my seeing him for some couple of months yet. There are also letter to be written to Nordic, Mrs. Reeves, Mr. Stephenson, and Mr. Torr. Phew!

I noticed the invasion plans of 'Dienst Aus Deutschland', which are indicated by three arrows in which the first is pointing from France to Hampshire, the second to East Anglia, and the third to Norway (massed air attacks) to the East Coast. If it comes off, within the next 10 days, as stated, it is going to be hot for someone!

This morning, Saturday, I chanced to be standing upon the Seacombe Landing Stage awaiting the arrival of the 'Wallasey'. I looked over in the direction of Liverpool, which appeared as a mist across the river, and the only visible thing was the Royal Liver Building. There is a little dome on the middle of the building, and it appeared at first as if it was expelling a red glow. This rapidly became more intense, and then it appeared as if flames were shooting from it. In another moment the top of a red capped the dome. It was a marvellous phenomenon. The simple explanation of this is, that the area of the dome was equal in area to that of the sun, causing a mock eclipse, which follows that of a description of the eclipse.

After reaching the other side of the water, and after I had viewed a battleship during the crossing, the 'Wallasey' turned to come broadside to the Liverpool Landing Stage. I waltzed over the side of disembarkation and noticed the top of the Cathedral Tower with its two deric cranes on top. The sun was again playing an important part in the scene. The only buildings visible through the mist was Royal Liver, Cunard, and Mersey Docks. Buildings in the foreground, and a few roofs disappearing in the distance. The Cathedral was enveloped in fog, with the exception of the top of the tower. The effect of the sun behind that upper portion through a red glow from the sides and top, bringing it out in a strong silhouette.

Having giving some impressions which may not be repeatable until next year possibly. I should be going on to attend a wedding at St. James, to sing in the choir – an occasion for two people that should only happen once in a lifetime!!

From your affectionate son,

Edwin

* * *

Letter 30

3, Ormiston Road,
New Brighton,
Cheshire
4th March, 1941

My Dear Mum,

I went for a walk to Leasowe on the first Sunday afternoon of this month. This point is not so far away as the one we reached in September last. It was a glorious afternoon. As I walked along the New Brighton Prom. The sun shone out of a cloudless sky, although a strong wind was blowing from the sea. Heavy waves pounded against the Prom. And threatened to bathe me at every few yards. I walked across the Leasowe Golf Course, on which a number of golfers were active. Spring was in the air that day. Some sheep with their new born lambs bounded and frolicked about the courses, and a few shy snow-drops blossoms peeped out from under the bushes bordering the Course. The little trip took just over three hours.

I have had a letter from Auntie Ruth again, who tells me that the Squarelys are ill – so they are keeping your two invalids company (metaphorically). I am trusting by now they are either in the act of recovering or have recovered. It must be difficult to manoeuvre them about at night in the raids, or do they just take a chance by staying in a warm bed.

In your letter of the 6th inst., you have one spelling mistake. It is certainly unusual for you to slip up in the orthographical field. But with so many things, mostly worries on your mind, it is almost natural. The word which struck my eyes while reading it was 'harassed', in the sentence, the correct spelling is: H A R A S S E D.

Musical activity is intensifying. My first lesson of this year was given on Saturday the 8th inst. Commencing from this Wednesday, the 12th I am to have a lesson each week at 6:30 p.m. each lesson provides me with ample work. Mr. Guest-Smith, F.R.C.O., L.R.A.M.is including HARMONY AND THEORY in the lessons. Perhaps you could send my Manuscript book, which is either in the top drawer on the left, or second drawer on the left of the bureau. I believe there are two together; the one with the lesser amount of work done in it is the one I require. In addition to the abundance of work in music at present, Mr. Guest Smith is anxious that I procure a second-hand copy of Mendelssohn's Organ Sonatas. I have already approached a Liverpudlian of the whereabouts of second hand music shops.

Last week I succeeded in finally checking and correcting my epistle to Stuart Durley. The 7-page foolscap effort was sent on its way Saturday afternoon, from New Brighton. It took nearly three weeks to compile.

I have heard most of the air raid news up to and including Saturday the 8th of March. Concerning the Home Town. The information was brought back by a member of the staff who had been on a week's leave. Since our warning and gunfire of Monday night we lived nine days without further activity. We then had a short-day raid on Wednesday afternoon and two more days, warnings followed the next day. On Sunday night we had two nights alerts, one with gunfire. On the following night (around Midnight) there were two alerts, but I knew nothing of them until I was told by fire-watchers in the morning. On Tuesday at lunch-time while waking outside Liverpool Cathedral there were two more warnings. That night we had the longest raid for many months which lasted a duration of 3¼ hours , I fell asleep before the A.C. from news of raids on other towns, it appears that Jerry is strengthening his attacks against us, and a glance at the tonnage of shipping lost shows more concentrated activity in that direction.

Wednesday night, the 12th, I went straight from the office to do some organ practice at St. James Church before my lesson. He was pleased. During the harmony lesson the warning had to *go, and he blamed me for it, and said that every time we have a lesson Jerry has to be present. The A.C. went just about dusk as we left the Church. I had arranged with the landlady previously to defer the evening meal until just before 8 o'clock. I, then commenced my variety program, and at within a quarter of an hour, another warning went and I closed down and transferred my musical attention to writing harmony. The moon had risen and everything predicted trouble. Jerry 'crates' started coming over and it was soon apparent (8:45 p.m.) that they were coming in waves, and a blitz would be certain. Continuing with harmony I heard other forms of 'ethereal harmony' blending with my efforts. Mr. Moody and I scrambled for the table, and down came the stick of beauties. Debris fell from the heavens like rain in a thunderstorm – I knew it was a near one. We retreated to the Hall under the stairs while the raid generally developed and stick after stick fell around. At midnight the lights fizzled out, and we substituted light with a candle. I had little sleep until 3 in the morning, but it soon slackened off, and I retired to bed as the A.C. sang out at 4 o'clock. Later the first parallel to Seabank Road, and joined by the short roads of Magazine, Manville, Ormiston, Onslow etc. at the top of our road there is a crater, and there is one in Manville.*

Thursday, the 13th, after a short sleep, it was time to be rise again for another day at the office. Today Mr. Storr, one of our Head Office Managers, was due to visit our Liverpool Offices. We took another circular tour of Wallasey to get to the Seacombe Ferry, passing many additional scenes of destruction. We found the Ferry to be closed, so took another bus to Birkenhead, where there were some lengthy queues. We joined one which leads us to a Liverpool Bus. This took us under the Mersey Tunnel. We arrived at the office half-an-hour late. There were one or two fires which sent smoke over the best part of the office area. It took 2 hours to get back home. We had out tea and went with the landlady's family to a big shelter in case of a repeat attack that night. We had been there for about 45 minutes before we were told the warning had gone. We could hear a great number of thuds, but being as the place is almost sound proof we heard less of the raid. We certainly felt more secure here than in a house. The A.C. went at 2:15 p.m., the next morning. We returned to No.

3 but could see nothing on the way to tell of much damage in our direction.

It's Friday and the ferry is still not running this morning and I have had to come in again by various buses. As there is so much infection around you I trust you will steer clear of the many dangers and maintain perfect health and safety.

From,

Edwin

***

Letter 31

3, Ormiston Road,
New Brighton,
Cheshire
18th March, 1941

My Dear Mum,

I am enclosing just a few lines with my soiled linen to let you know mainly I am safe and well.

Music was forced out of my life for 4 days by taking the necessary precautions of taking shelter each night in case of repeat attacks, which came nevertheless. I do not feel so prepared for this week's music lesson as I did last time, and feel that there is little time left to work.

Despite adverse conditions for cooking, I find the landlady has maintained an even standard of efficiency throughout the many hardships and inconveniences of the past week.

My weekly letter should contain a fuller report of the Blitzen. In that report I am only endeavouring to give you a general idea of the new conditions under which we live. I am endeavouring not to mention towns. I have heard also that you have not escaped from a number of raids, but I still trust nothing has come your way. Last Saturday afternoon was the last alert we have had, and it is now Tuesday.

I have now been back for 4 ½ weeks, which is either half-way or a little more. However, it must be half term!

From your affectionate Son,

***

Letter 32

> 3, Ormiston Road,
> New Brighton,
> Cheshire
> 22nd March, 1941

My Dear Mum,

Once again, I thank you for your parcel of Laundry and letter which I received on Friday, the 21st inst. It was a coincidence that I should buy some chocolate Friday morning, and to come home and find you had sent some. Your parcel took four days to reach its destination. I expect by now you have received mine, which left me on Tuesday.

As far as the Heads of my Department are concerned, they are willing for me to go on weekend leave next weekend, commencing the journey on the 27th.

I am due for a week's leave this year, and I have already been asked when I should like to take this. I have left the question in abeyance until I have discussed the matter with you and my father. Perhaps this year, we may be able to have our holidays together. A member of our department has taken one week's leave together with weekend leave, this weekend, which together makes a total of 11 days. This would, of course, apply in my case.

Further evacuation schemes come into operation next week. The authorities have been very slow to realize their perilous position. Mrs. Wills' two boys are being sent away next week to Oswestry, and she is probably going with them. In the event of her so doing, it will be necessary for me to fond other digs. To prepare myself for the change, I am going to search one or two districts of Liverpool for suitable digs. As I have already mentioned to you in January last, when I was on my 4th weekend leave, I had the district of Childwell in view.

In view of the fact that I am coming home in the very near future, all being well, I am not venturing to send the report I mentioned in the parcel, but will make another statement, more detailed that the one prepared for the Post, and I hope to bring it with me.

Music is at a standstill, and it looks as though it will remain so for a long time. If I go to the other side of the River I will of course cease to carry on lessons at St. James, New Brighton, and in view of the fact that it will not be many months before I will be eligible to register for Service.

I am cutting a new tooth, which is bad enough in itself without the additional condition of recovering from 'strafing'. The first came through the gum on Wednesday, a week after the first Blitz, and today, Saturday, a good deal of the tooth is visible. I take it, that it does or not I cannot say, but leave others to form their own opinion.

You are having a worrying time of it, with Jerry coming over night after night, leaving you in suspense as to whether he is going to turn his affections to you, or whether he is going to pass on. We are fortunate in that way here. Of Jerry comes over, we know we are for it, that is, if he comes early in the evening.

I once more trust that you will keep safe, and that we will have nothing more of a serious nature, until I should arrive home – early Friday morning the 28th.

From Your Affectionate son,

Edwin.

***

Letter 33

3, Ormiston Road,
New Brighton,
Cheshire
18th March, 1941

Dear Mum,

I enjoyed the comfort of the Southern Railway coaches for an hour and a quarter, until I reached Basingstoke, where the 10. Train was backing into the fifth platform. This is never lighted, so I sit in the dark for 30 minutes until I reach Reading. I arrived at the latter station at 10–25 where I had to wait until 1–10 for Birkenhead train from Paddington. While waiting on the station I met one of our colleagues from New Milton, who accompanied me that time we were stuck in London. I managed to get a little sleep while waiting in the waiting room. Shortly after, the crowded train arrived, and after much difficulty I found a seat. I was able to sleep until just after dawn, when I reached Ruabon. I believe I dozed again but I am rather hazy about it, the next station I saw was Chester.

The train arrived early at Woodside Station, Birkenhead for at 8.50 a.m. I was walking along the platform in the direction of the Buffet, where ate a cake, and some biscuits followed with a cup of tea. The sun shone brightly as I crossed the River, and appears symbols of the prospective weather for my seventh term.

From your affectionate son.

* * *

Letter 34

*3, Ormiston Road,
New Brighton,
Cheshire
18th March, 1941*

Dear Mum,

I did not send my usual weekend letter, until now as I wanted to include the results of the proposed visit to Childwell. The letter I prepared has been left over at the above address, hence I must do my best to draft a fresh one.

I thank you for your two parcels, the first I received on Wednesday, and the other received on Saturday afternoon. As usual you are continuing with your four-a-day raids, which still gives me the strong feeling that you are on the Front Line in the Battle of Britain.

I visited the Cathedral as planed when I was home, and found the building still whole. Mr. Goss-Custard did not happen to be practicing that dinner time so I did not stay long.

The landlady has approached me on two occasions since my return regarding the matter of finding other accommodation. There are two reasons for her doing so. She considers N.B. as unsafe area, and for my safety, desires me to leave it. She is taking a 'holiday' this week, as an experimental evacuation which, if successful, will turn into a permanent evacuation after Eater. In which case I have to find digs at least for a week.

In accordance with the desire to remove from Wallasey and in satisfaction of exploring (and as mentioned when I was down) I set off for Childwall, Liverpool, 16, on Saturday afternoon and in 25 minutes I arrived in the District of Old Swan. I had spent best part of the morning studying maps of the area I proposed to 'invade'. After passing through Lord Street, passing dirty buildings all the way, the car passed the St. George, and soon entered the long road leading to Old Swan. I soon sighted my landmarks and found Queen's Drive, which was to lead to Childwall Five Ways. From here I took a direct route to Childwall Abbey, set in rural surroundings, which seem quite opposite to those of the rest of Liverpool. The door of the church stood open, and I passed through its portals. The organ was playing (rather noisily) as I walked around the rather cramped place. Although it is an Abbey, it is no bigger than Holy Rood Church (that was!)[3].

I returned along Childwall Abbey Road, which led straight into Dunbabin Road, the road given me by Mr. Sims. The numbers decreased as I walked down the road until I came to 91. I am rather perplexed at the result of knocking, as they do not know and have never known anyone by the name of Sims. Therefore, at the moment, I have nothing in view. But I am glad I have seen Childwall, with its classic and modern houses and its marvellous main highway joining Childwall, Broad Green and following round to Bootle. This highway is divided in two but a bank of grass and trees in the middle, leaving the up traffic on one side and the down traffic on the other side. On the way back, I called at a modern Church. This was called St. David's, the organ has already been installed in the North Transept.

I arrived 'home' just before 5 o'clock, making the journey a total of 5 hours. The time taken in buses, ferry and buses on the Wallasey side of the River, starting from Broad Green Road Liverpool, 14, was 1 hour and 30 minutes.

Up till last night, the 6th *I had heard no alerts. There was one early one morning, after midnight, but I did not hear it. However, last night, as before stated, I heard a warning, which woke me up at 11:30 p.m. I did not go to sleep, so sat up and at 12:15 a place came suddenly, and a terrific barrage was put up. I retired to the Ground floor and remained there until 2:30 a.m. after hearing many several bursts of gunfire. I did not hear any bombs drop. Again, today we had another warning in the daytime. Jerry is warming up for another Blitz for us.*

I am pleased to see the Italian Navy going to the plane where they ought to go – Davy Jones! In the absence of this fleet it will leave |Admiral Cunningham and his fleet without a rival and ruler of the Mediterranean Seas. The Yugo-Slav Situation has reached its peak, on Sunday morning at 2:00 a.m., when Germany declared war on that country and its neighbour Greece. It is here that I hope to see tough resistance and the path to final victory on land, in the sea, and in the air.

---

[3] Church in Central Southampton which was destroyed in the blitz of Southampton a year before.

During the past week I have suddenly realised the fact that in about 2 months I shall be well within the registering limit and to enable me to have a better choice I suggest volunteering. I will obtain particulars from the RAF sometime in the near future.

I do not yet know how much Easter Holidays has been allocated to us, but I expect Monday off. On Sunday I hope to go to the Cathedral service at Liverpool as I hear the service is highly commendable and is often broadcast over the radio. If there is time I hope to see Chester Cathedral and the City.

You will not forget to forward the camera as I hope to find something that I can show for my travelling.

From your affectionate son,

Edwin

* * *

Letter 35

New Brighton,
Cheshire
Good Friday, 1941

My Dear Mum,

Another week has passed. Today finds me staying at No 10 Ormiston Road, of the above town. It is very still; the sun is obscured and yet it is not dull. Although there are no brilliant sunbeams and it is very mild, it is even drizzling now for a minute or so.

I am alone with my thoughts. The landlady and her younger boy have gone to Nantwich to join Stanley at his billet, just for the Easter Holidays. Mr. Will is working most of Easter. I still sleep at No. 3, and I have Breakfast there. Chas. has gone to Southampton for 10 days. The occupants are No. 10, where I am writing this letter this afternoon, are the mother and father of the landlady's husband at No.3. Naturally, they are an elderly couple, and are very homely. both are from Southern England; one from London, and the other from Devon.

Raid news continuing from my last comments on my letter of the 7th, goes on as follows; Monday night was 'hot'. The alert sounded at 9:30 in the evening, and 45 minutes later, the first planes dashed over, not so directly overhead as in raids hitherto. These, as well as scores of others were met with intense AA fire, which for whole minutes blotted out any other sounds. The raids lasted for 7 hours with only a few breaks. A number of 'unbroken Easter eggs' were lying about next morning, none of which were near here. A few H.E.s were dropped at Old Swan (neighbouring district to Childwall), through which I had passed on Saturday afternoon during my excursion.

This morning, I took a short walk to Liscard with C.A.R. I discovered even more damage, done in the March Blitzern. It was caused by another mine. The end of a road and several adjoining streets are scarcely recognisable as such. A motor shovel crane is at work shovelling up the piles of wreckage, and remains of houses. I visited St. Mary's, Liscard, where I saw the organ. This will be another for my collection.

I have had dinner at No. 10 today, and when I want to practice I go over the road. I have spent some time at music today. Part of this afternoon I practiced at St. James.

On Tuesday and Thursday of this week I have visited the Cathedral at Liverpool. Each time it has been during the lunchtime. On the former occasion I did not have the chance to hear the organ but on Thursday just as I was leaving I heard the heavy footsteps of the Organist approaching from the Entrance and so I stayed on. He started by doing chromatic scales on various stops, and with his marvellous extemporizing ability developed many melodies and harmonies from them. I stood for some time watching and suddenly I saw all the stops on the right-hand side fly out and – you have never heard anything like it. The cathedral was filled with music; vibrating, echoing, and re-echoing, the great tones from that king of organs was released and they spread out across the entire length and breadth of that domain. Unfortunately, before he finished I had to fly back to the office.

Tomorrow I hope to visit Chester; where I will spend the whole day, if it is fine. I am getting quite excited about it.

On Sunday, I hope to go to Liverpool Cathedral. Trusting all is well and will remain so.

From your affectionate Son,

Edwin

***

Letter 36

3, Ormiston Road,
New Brighton,
Cheshire
18th April, 1941

My Dear Mum,

Again, I thank you for another letter, dated 15th, and received by me this morning. But I am sorry to hear your foot has met with an accident, and hoping that it will quickly return to normal.

Members of staff have returned from the weekend leave have released information regarding 16 mines which were dropped at home. I have also heard that a plane was brought down near you.

For me it was my first Easter away from home; the second festival season celebrated in the North. For a long time, I had been waiting for the opportunity to tour the City of Chester, and with a few days to spare during Easter, I seized one of them for the purpose.

I woke with a jerk on the Saturday morning, which put a kick behind my steps and movements of the day. I had had a refreshing sleep and felt ready for the adventure. There was a cloudy sky, but fortunately it was not raining. I had my breakfast at No. 10, and at 9:25 a.m., I jumped on board the New Ferry Bus, which for the small sum of 3d. took me into Birkenhead, as far as Woodside Station. Just by the station is Crosville Bus parking ground, a service of buses which serves the whole of Northern England. I found a double decker bus waiting, marked 'Chester', and I secured a front seat position on the top. It was not long before we (fellow passengers and myself and, of course, the driver and conductor!) were leaving the grimy building muddle of Birkenhead behind us. We were soon running along the New Chester Road, a fairly clean, and straight highway. I can well imagine that its original foundations were laid by the Romans.

The scenery was mainly flat, but dotted with trees and stretches of newly ploughed fields.

We passed through many little villages, with their little old churches laying aside from the road, nestling amongst some beautiful trees. It seemed so peaceful in this country district, but here and there, holes in the fields, and shattered farm houses painfully remind us of the Battle of Britain. Occasionally, the River Mersey came into view across the landscape.

After a 95-minute journey in this conveyance I arrived in the old market place at Chester, Northgate Street. I walked away from the bus after 'photographing' an impression of the building near the Bus stop, in order that I should recognize the place when I should require the bus to return. Then slowly walking away, within only a few feet, I turned a corner, and there was the Cathedral. This was a truly ancient structure, built of reddish-brown stone hewn in the area. I made this my initial, and chief objective of the visit. I entered by the door in the South Transept, and looking ahead I saw the gilt pipes of the Organ. The instrument is placed in the Nave in front of the Choir Screen, and entrance to the North Transept. I found that the loft is supported on three arches on double marble pillars. The organ is place in a beautifully ornamented case, with a decorative screen around the front carved to harmonise with it. There are two other sections to the organ, another is on the South side of the choir stalls; and another section is in the North Transept. The latter section contains four of the Pedal stops (including the thirty-two foot reed and the thirty-two foot open wood).

I obtained a guide book in the Cathedral, and read through its pages as I sat in the nave. I chanced to look up and noticed two heads showing above the carved wood surrounding the organ loft. My attention immediately focused on the Organ. Several groups of people had formed in the South Aisle near the Transept, and by their dress I guessed a wedding was coming. The organ pealed forth, and all the while I endeavoured to classify the instrument. The organist rushed through the Bridal Chorus and the wedding march. After the ceremony I recommenced an inspection of the many interesting features of the Cathedral. In the early times this old building was an Abbey, and in the North Archways are still to be seen. In the South-West corner is the foundation for a tower, which was never built, *owing to much controversy. The Transepts are very unequal; the North is small and the South extremely long – so much that on entering for the first time I thought it was the Nave. Some years ago, this was converted into the Parish Church of St. Oswald's. the Nave of the Abbey, too, was once upon a time separated from the choir and the Chancel by a thick stone wall, for a similar purpose. The old monks being superstitions, thought the Devil peeped at over this wall while they were praying and chanting in the Choir. To frighten him away, they carved a stone image of a devil in chains, and placed it on the other side of the wall. Although the wall has now been removed the devil is still to be seen, his position now being in one of the window sills in the Nave. The Cathedral is rather oddly shaped with many unexpected corners and nooks, which perhaps creates a greater interest in the place.*

From the Cathedral I made my way along the busy thoroughfare of the city. There are four main streets arranged in a square around the City, names after the four corners of the Earth, Northgate, Eastgate, Westgate etc. the city is bounded by the old walls, which are in a state of perfect preservation. It is possible to walk round the entire city on the path along the top of these walls. I went for some distance and eventually crossed the river Dee, which is outside the Walls. This bridge is narrow, and short, as the Dee at this point is not wide. I visited the Parish Church of St. Mary's on the Hill Without the Walls. I returned from here and re-crossed the Dee and discovered the parish church of St. Mary, Holy Trinity Church and St. Peters Church. The oldest was St. Peters as most of the original structure of pillars were still there. Each of these old streets are bounded on either side by Tudor buildings, many of which have slipped into odd shapes. the streets are narrow and many of the buildings protrude over the pavements, supported by columns. It is curious to note that these shops, cafes, and inns have two fronts! And they have two pavements each side of the street! When you stand on the pavement on the road level you are standing under the other pavement which runs along over a colonnade at first floor level. These upper pavements are bounded by balustrades, and connected by bridges over the street at various intervals.

Later in the day I returned to the Cathedral to see the console of the Organ. A thorough inspection of the organ was impossible as Evensong was just about to begin. I stayed to the last feasible minute at the Evensong, until I had just enough time to walk back to the bus stop. At 5 o'clock, I departed for Birkenhead, leaving behind me the old world city. The journey was finished at 6.30 when I had a good tea at No. 10.

On Sunday morning, I fulfilled my pledge to go to the Cathedral. The service was so well attended that within 30 minutes of its beginning all seats were taken, save three. These were near the front at the extreme left corner. The position was by no means favourable, for seeing the moments of the Choir and Clergy. I found myself looking straight down the North side aisle, with a gigantic pillar, and the pulpit obscured the chancel from my vision. I also had a lopsided hearing of the organ, as the great 32' pipes, and 64' of the Pedal Organ were overhead, and as they were much in use during the Festival Service. I could not hear anything other than the thundering of the 'Fundamental Bass' of all the chords in creation.

Mr. Goss-Custard had already commenced playing before I had arrived, so had a little recital before the service. The Te Deum was one of Vaughan Williams, which included many unaccompanied parts. The anthem was taken from Beethoven's 'Mount of Olives' and was delightfully, and enthusiastically sung by the choir. At there was a celebration of Holy Communion after the service there was only a very short voluntary. After having a fine lunch at No. 10, I did a fair quantity of music in the afternoon at No. 3, and after having tea I went to St. James for the Evensong. This was the first time I have attended a service here before some time, owing to being away on leave one week and visiting other churches in the meantime.

Having finished the service, and a desire to brighten things up, I took a short walk around the Tower Amusement Park, with its Switchbacks and Flipflaps, Autocars, Rollerskating etc. here the people of this district were making full use of the time to 'Banish the blues' as the maxim of the Tower says – and their money! – It was a treat to see life, as of late this area has become a ghost town.

On Monday, I did a great deal of music in the morning, both on the Piano and on the organ. In the afternoon crossed the River to attend a short organ recital at Liverpool Cathedral. There were not many present, and the majority were in uniform.

Letter 37

3, Ormiston Road,
New Brighton,
Cheshire
27th April, 1941

Dear Mum,

Thanking you very much for your parcel which I received yesterday. It proved very interesting.

The nosegay has not yet revived although I have doused it with plenty of water. The garden must be looking spring like now with the daffodils blooming and all the other flowers coming on. The only flowers I see here are those in Woolworths, Liverpool. I have seen some poor specimens of pansies (I take it that is what they are supposed to be as they have no names and just marked 3d). There are a number of other flowers which is a nice sight. Hence, every time I go to 'Woollies' I visit the flower garden.

You have been having several raids lately. CHM came back from leave full of it, and fed up with the nights in the shelter. Even here we have had visits most nights. Who said Jerry was weak? He has been pasting a town a night for well over 3 weeks now, it is continuing to do so. It must not be forgotten that he is also maintaining a full-scale force in the East. It stands to reason that each new country he gains, he extracts additional war materials from them. For instance, whereas France possesses some of the largest munition factories in the world; these until Spring of last year produced munitions for our Allies, since then these factories have produced supplies for German use. So, it is with the other countries. It is now that we are beginning to feel the result of this overwhelming stock. I fail to see the use of our blockade as a paralysing force; it is little more than a nuisance to him.

In connection with this, I saw a very good cartoon about 10 days ago, depicting a man sitting on an island reading a paper called 'Sunshine stories' it said 'Germany short of oil', 'Germany short of food'. In the corner of the picture a heavy cloud was drawn – the gathering storm' was written in it, to elucidate its meaning.

On Saturday (19th) I made another trip in Liverpool, this time in Wavertree. The less I say about it in the letter.

Late night, Saturday (26th) the warning sounded shortly after I had gone to bed I dressed quickly as I felt there would be trouble as the alert sounded early. We usually have a precedent during the day but this time there had been no day alert. Within a few minutes the hounds came over, flying below the cloud. We had not time to go to our usual shelter, so made do with a place over the road. For 3½ hours Jerry did his worse – but fortunately not near us. Colossal echoing explosions indicated that it was the City. I have not heard any official news yet today – Monday will tell! It is possible we may have a repeat blitz tonight.

Thank dad for his good sketch of the garden. Regarding Mr. Small, we should have heard if anything has befallen him, as our agencies forward all such information to us by cable. It is forbidden to mention movements of vessels in correspondence. I am only at liberty to say that he should be home shortly.

Trusting everything is well, and that your foot is even better.

From your affectionate son.

***

Letter 38 (postcard)

> 3, Ormiston Road,
> New Brighton,
> Cheshire
> 7th May, 1941

Thanking you for your welcome letter received Monday. I have had little sleep for the last 6 nights. The letter I sent you on Saturday will be the last sent from that building; it was gutted that night. I cannot describe here how things look but you know – only worse.

I have to report every morning at 10:00 a.m., in a building which our Co. have used pro-temp as a rendezvous. Our manager has gone to Head Office to report the situation and our position.

***

Letter 39

> 3, Ormiston Road,
> New Brighton,

My Dear Mum,

> Cheshire

It is with great difficulty and reluctance that I endeavour to recollect the events of the past six days and worst of all the past six nights. Owing to an incredible loss of sleep I seem to have spent the majority of that time in a trance. By now you will have received my letter dated 3rd, and my card dated 7th. The former will have given you a slight idea of the situation up to Sat. morning, and the latter only intended as a communication to relieve your mind of any anxiety created by news in the papers. This is intended to give you a better idea of the situation as it stands today, and also a brief summary of these strafing's. I find it slow work handwriting my private mail which seems to have accumulated all of a sudden.

However, on Saturday night the warning sounded even earlier. Ah! That night I had not even a lull of 5 minutes to doze. It was far beyond imagination. For close on 6 hours I felt literally thousands of detonations. However, at 5 the next there was at last a cessation of the almost continuous thump, thump, thump! Then

I ventured from my hiding place and out into the open. I was met with a brilliance equivalent to a glorious sunset, at a time before sun has commenced, I turned towards the Liver and I was astonished at the sight which met my rather unsteady gaze. I moved to the top of a bank overlooking the prom, which gave me the opportunity to see the most vivid sight of a lifetime. Across the River there appeared to be one enormous fire stretching in each direction as far as I could see – covering a distance of 6 miles.

I had been a clear night, until great volumes of smoke commenced to roll skywards, and they reflected the great inferno. I was still horrified at the sight, when an explosion occurred in a building that was well alight. It resulted in sending showers of sparks into the 'clouds', like a firework display. Then another went further along, and a few seconds later, another. Then the long-awaited A. C. sounded but there was only one siren – usually all the sirens go together here. Everyone relaxed. As they came, one by one into the open I watched their reactions to the sight that met them; I noted their horror, their surprise, and their pessimism! I am sorry to say that scene did little to support our confidence in the final issue.

I was soon in conversation with many of these bewildered gazers 'who said he hasn't the stuff?' they questioned. Others who have seen raids on other large cities, including Londoners, people from Plymouth, etc. said they had seen nothing to compare with this. I retired from the scene half stunned by the memory of it, and more or less staggered home, tired and exhausted. I laid on the bed at 5:30 a.m. and tried to sleep. Then another AC wentan explosion followed. So, I commenced timing these explosions. They were at intervals of approximately every 2 to 3 minutes. But I fell asleep. I was awakened at 9 by a terrific explosion, but soon fell asleep again.

In the afternoon I went across to see the damage. Only few of the fires had been quenched by 2:00 p.m. I saw fire engines from Blackpool, and even Coventry, assisting the auxiliary forces. The brilliance of their work got all but about six fires under control before nightfall. Again, that night we endured another attack nearly as fierce as the preceding night. This started many fresh fires and when the AC went at 4.30 it looked very nearly as bad again.

Now it was Monday. I went across the river as usual with other office workers. Many large fires were still burning. I felt fed up only having 9 hours sleep in 4 nights. The front entrance is the only part of our buildings untouched by fire, and there we stood among 2000 other of the employees and tenants of that building. The floors of the place were still standing. Many of the inside corridor walls had collapsed exposing rusted girders. We made excursions to our various rooms, now shells, for salvage purposes. It was hot as the walls retained some of the heat. In our 'office', the Victualing Dept., was not recognisable. The only articles remaining were: one safe and 2 pieces of rusted metal which had been two typewriters, 2 days previously.

The manager set off for Head Offices, and we were told to report at Marwood and Robertson's, one office left on the other side of Water St. having the rest of the day off I went back to bed in the afternoon.

The next two raids, Monday and Tuesday did not commence so early, in fact not until about 12:30 a.m. hence the newspapers say 'raids on Merseyside early this morning'. Both raids caused several fires again, one on this side of the river on Tuesday. On Wednesday some of our departments have temporarily settled in Cunard Bldg.; while we have gone to Royal Liver Bldg., in the Canada Pacific Office. Apparently, we will still remain here.

I should be due for weekend leave shortly if they permit.

From your affectionate son,

* * *

Letter 40

3, Ormiston Road,
New Brighton,
Cheshire
12th May, 1941

Dear Mum,

I arrived safely at Birkenhead at 8:50 a.m., Monday, a little earlier than usual. This gave me a time to spare before reporting at the office. Our offices are still at the Royal Liver buildings. Again, I have the rest of the day off as has been the case during the last week. Apparently only slight damage was done during my absence, although they have had several night alerts. There was one at 8:00 a.m. this morning before I had finished the journey.

It has rained this morning and still looks that way inclined. I am feeling much better for the weekend and have already started looking forward to my next visit. My gland has ceased to bother, although one large ulcer still remains at the back of my mouth.

Trusting there will be no other bombs nearer than those of Friday night for some time. My mind is still focused on your garden and the animal life at home and down the lane. Those wee kittens were just too sweet.

From your affectionate son,

* * *

Letter 41

3, Ormiston Road,
New Brighton,
Cheshire
16th May, 1941

Dear Mum,

As yet there has only been one warning since my return, apart from the one which sounded as I arrived. That other warning went on Wednesday afternoon as I was eating your mince pie. I have had three good night sleeps, each representing 9 hours.

Our Department are still using the Royal Lier Building as a rendezvous, where a CPR counter serves as a writing desk for me. As we have no typewriters we have to hand write the little correspondence which we do receive. That's for 2 hours a day I spend here, returning at midday to the digs.

On Monday afternoon, I spent on the piano until Mrs. Wills returned from Nantwich and Rodney engaged more of my attention. Tuesday afternoon, I utilised my first rise which should mature at the end of the month. I deduce that a slight increase will be noticeable in this month's cheque. On Wednesday afternoon I ventured for the first tie for 10 days to penetrate further into Liverpool than the Pier Head. I observe that many of the roads are impassable. Exchange Station has been seriously damaged by a land mine and Lime St. Station is not so devastated. Lewis' great store is gutted as is all the other big stores around it. Lord Street and Castle Street are now lie our High Street, many other streets look like our London Road. At. Luke's and the end of Bold Street are gutted. The Cathedral still stands surrounded by eight neat craters. One bomb fell on a corner house opposite, another on the Cathedral book shop; one in the road, toe of the Cathedral steps, another fell outside some swellings, while yet another found a target among the gravestones. The works situated on the die of the Nave has been gutted, leaving twisted cranes etc.

On Thursday afternoon, I took myself for a walk in the brilliant sunshine, which suddenly burst out, the first since my return.

The landlady is again going away to N. this weekend and has arranged for me to stay for 3 days in Seabank Road. It happens to be a place where another four staff had stayed for 5 months.

Last night, we had another warning, but it failed to rouse me. I heard it at 2:45 a.m. but do not remember hearing an A.C. I am told the alert lasted about an hour and a half.

I notice that the book I am reading is addressed to my father, dated 11th May 1918. It was the 11th May 1941 that I brought it here. By the way has my father read it?

Hoping everything remains quiet at both ends.

From your affectionate son,

Edwin

P.S. Your letter and parcel had arrived together on Friday, 9th.

***

Letter 42

3, Ormiston Road,
New Brighton,
Cheshire
24th May, 1941

Dear Mum,

Thanking you for your two letters, the first received on the 19th and the second on Saturday the 24th. Usually in the past your weekly letters arrived on Friday mornings; then my weekly letter written on Friday contained the reply and events of the week. Of late, your weekly letter has not arrived until Saturday, or even the following Monday making my 'home correspondence day' one day later.

Last weekend, from Friday evening until Monday morning (16th–19th) was spent at 287, Seabank Road, just around the corner from my original digs. It is the same type of house, large, but crammed with an older type of furniture. Much of the furniture being lately removed to the ground floor. Most of the time I spent reading David Grieve, more or less to create the impression that I had something to do, apart from my interest in the story I hate people to think I look lost. On the Saturday night, the landlady of this abode, Mrs. Halsall, of about 60 years of age, and her daughter of about 30, were going to the cinema at Liscard. I went with them to see 'Freedom Radio' at the Capitol. On Sunday I went to both services at St. James where I enjoyed singing. Late on Monday afternoon Mrs. Wills returned and took charge of me again. Your letter had arrived that day, and immediately dealt with it.

The Hess situation has baffled me as much as it did everyone else. Why should he, Hitler's deputy (of all ranks) come to the enemy country? Why did he not go to Spain, Sweden or another neutral country like USA? However, I have a few ideas. His disability, Schizophrenia, or rather ability, as his Fatherland try to make out I can understand his mentality; one part of his brain thrusts after power, ritual, and vanity: and the other part, searches for peace, religion, and friendliness. These two minds alternate – at the moment he is in the latter frame of mind, then after a short transition period, his brain will have switched over to the other more brutal half – then we may have something sensational!

Owing, perhaps to rather bad weather conditions of the past three days Jerry's aerial activity has ceased, but up till then, we had warnings each night for the past fortnight, with only one or perhaps two exceptions. The first half of this week brought perfect spring, or more like summer weather to this Northern seaside resort – New Brighton. As I have only been working two or three hours each day commencing at 10:00 a.m. in the morning, I have plenty of spare time to be in the sun and by the sea. It has been very like a holiday.

The office may soon be settling down in Martin's Bank Building, a very fine edifice, better in refinement and design and material than India Buildings. It is a large, and from the exterior it shows much more elaborate ornamentation. It is nearly opposite India Buildings as it is in Water Street. Although Liverpool is not really worth seeing in either its present or preceding state, I should very much like you to come one day to see where I have been working for the past 10 months.

The book has been a great help to my spare time. I have read the first two books, Childhood and Youth and have commenced the third book Storm and stress, and have reached the second chapter, page 153 David, since my last letter, has run away from his Aunt Hannah, and has been four years in Manchester.

My weeks holiday is, to all intents and purposes, quickly drawing near. The actual holiday is due to commence, according to the list on the 7th June, Saturday, but I may be able to leave here on the 5th, Thursday, arriving in the same way as I do when I come for a weekend. I will be due to report at the office on Monday, the 16th of June, leaving Southampton, on Sunday 15th.

So, I will close now, trusting things will remain fairly quiet – it is about 12 days, before I may be home again!

From your affectionate Son,

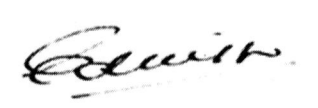

***

Letter 43

3, Ormiston Road,
New Brighton,
Cheshire
29th May, 1941

Dear Mum,
I am enclosing my soiled linen which has accumulated during the past fortnight. I can only find three collars for the two shirts, but I hope to find the other one.

The weather has been full of thunder this week. On Tuesday and Wednesday, we had a short storm early in each evening. But, on the latter occasion the weather conditions did not prevent an air attack.

I hope to write my usual week end letter – when I can find the time – as I am now BUSY. By the way Mr. Small has not probably come home, but according to the present conditions, the last day of this month should see him somewhere near his home. Whether he will be permitted to have leave is a matter about which I have little idea. But that is as much as I know. Mr. Moody is again on a week's leave this week, but mine should be coming next week.

From,

***

Letter 44

3, Ormiston Road,
New Brighton,
Cheshire
30th May, 1941

Dear Mum,

I thank you for your long and interesting letter which I received this morning, Friday – the usual home letter day. Yes, I well remember our trip to Portsmouth, I believe it was 1928 or 1929, when we saw the 'Hood'. I told my landlady I had been aboard that battleship, but could tell her nothing about it as it goes too far back into the abyss of my memory.

At 1 pm on Monday 26th May, our department took possession of first class accommodation in this great and magnificent building known as – Martin's Bank Building. It is almost opposite the blackened pile of India Buildings, in Water Street. You might be interested to know that a bomb had fallen at the rear of theses premises during the great Blitzen of early May. The damage sustained to this building is slight, and it does in no way affect us on the 4th floor. As you can see, by this script, I have scrounged a fine typewriter which I selected from 25 of them sent from our Head Office, but you may guess I had to fight for it! It is an Oliver, No. 20. I have also obtained another typewriter for this department, which is an underwood. Previously we had Imperials in India Building, which we brought with us from Southampton.

Hence the vast stock of work which has been slowly accumulating (the siren is just sounding 1:20 p.m., we shall get it in the neck tonight) – sorry for the interruption in what I was saying – but the work which has been accumulating is engaging every minute of my office life. It not only includes duplicating stacks of charred papers, but also the work of the past month. In addition, there is the prospect of something coming in very shortly, which will augment the work. As a result, I regret to sea, the position of my week's eave is in a precarious position, and will, I fear be liable for postponement. Perhaps, you could tell Auntie Ruth, if she is near, to put her wise.

Last Wednesday night, we had a short and sharp raid during which a number of high explosives and large incendiaries were in a residential district of the City. But now, you see in the paragraph preceding that thee may be another tonight.

I must have known what you were writing yesterday when I put my note in your parcel regarding Mr. Small. That I think will clear the position. But again, I should add that nothing is definite.

Yes, I am sorry about Mr. Gadd, he was a fine musician. I believe he has been ill for some time – even when I saw him in Feb 1940 he had teeth trouble.

You did have a thrilling time at the Luncheon – but you should have proceeded with your lunch. I am sure the Lord Mayor of Southampton did not intend to put an end to your Dinner! And again, you were very lucky in the Savings Certificate Draw.

Your life at that end has been quite the opposite to mine. To generalise it something like this – stress, speed and daring during office hours; patience (at cards) and book reading (David Grieve) in the evenings with few acquaintances with which to pass the time away. I have reached Chapter XIII.

Trusting you are still well and safe, and that circumstances will remain favourable.

From your affectionate son,

Edwin.

P.S. My leave has been postponed. I cannot come now until the pressure of work eases – goodness knows when that will be.

***

Letter 45

3, Ormiston Road,
New Brighton,
Cheshire
6th June, 1941

My Dear Mum,

Thanking you very much for the parcel dated the 3rd inst, received yesterday morning, before I left for the office. I tore it open to get at the correspondence within. I have been waiting more or less all this week for a shirt and collars but now the one received yesterday will do for next week. I also thank you for the chocolate.

Referring to my last letter of the last Friday, if you care to re-read the 'interrupted paragraph', which I typed while the day alert was still sounding in my ears, you will see my prophecy for a disturbance. My prophecy was materialised at 12:55 a.m. on Saturday, and went to the shelter in the back garden, a very limited area, with the landlady and her husband. Many planes were engaged in the attack, which was fierce for an hour or so. On this side of the water, the seaside resort in the address at the top of this letter, was apparently the chief town for attack. As the heavily laden planes droned low, I heard the beginning of the whistles in the distance and for the next 15 to 20 seconds during which the bombs made their downward course, those whistles became louder in an even crescendo until one was terminated with a crash followed by five thuds, but no explosion. Incendiaries were dropped but no fires were started. The landlady then decided to depart for a week, leaving for Nantwich for the next day.

That decision opened an opportunity for me to do the same. The last of the planes did not leave us much before 3 o'clock, and the A.C. followed at 3:30 a.m. In the morning I had a haircut, and thence to discover the vicinity affected by the previous night's raid. Within 5 minutes' walk from No 3 I reached a point in Sudworth Road, where one bomb had demolished the rear of two houses. There were no casualties. The 5 thuds turned out to be 5-time bombs in a line, continuing from the one which exploded on contact, along Glen Park Road. The incendiaries had fallen within 10 minutes' walk from my base, and buried in sand from the beach by fire watchers. I then made carious calls on landladies. Eventually, Heswall was the village for attack, and I bused to Birkenhead at 11 o'clock and from thence in a Crosville Bus to Heswall, arriving at just past noon. Heswall is about 10 miles from New Brighton.

The first person I was to call upon was a Mrs. Amery. On arriving at the Heswall Bus Station, I enquired how to get to Whitfield Road. The Inspector who I asked happened to be Mrs. Amery's son. He soon put me on the way. It seemed very promising up to the present. I soon found the place, but there were no vacancies. Then a chain of attacks followed. I was recommended from here to a Mrs. Penderville, then on, to a Mrs. Ford, and on and on again, each time meeting with the same reply, although sincere and polite, but still having no success, as regards to the object of the visits. I passed a fried fish shop, and the smell of the contents of that catering establishment struck me forcibly at the time (2 o'clock) when my inside began to call for a little attention. I entered and after satisfying my immediate requirements began issuing suggestions that I was on the lookout for a room for one week in Heswall. I again met with no success in the conversation which followed, but learned that some bombs had been dropped in this place the night previous. I went on, and presently I came across a bombed garage – and that was what these Heswallites were so indignant about – their first 'packet'.

It was a brilliant day, and the beautiful lanes, bounded by slopes of red earth and rocks, covered with bracken, pine trees and ferns, and other natural botany peculiar to Devon did much to please my senses.

I finally made enquiries at the Police Station, and as this organisation could not assist me any further, I decided to drop the matter. I returned late in the afternoon to settle down once again for a towing in New Brighton. The landlady had by now set off for Nantwich, and had left instructions at 'Grandmas' (No. 10) to take charge of me for a week, should I fail in the Neswall excursion. You will remember that I stayed here before at Easter; it is rather singular that it should occur again at the next Holiday Season Whitsun.

That Saturday night I went to bed again half-dressed, in preparedness for the third attack of the week. It came, a quarter of an hour earlier, and I awakened by the old man Wills. With six of us in their shelter, the elderly man, as usual, continued to narrate unceasingly in true 'Uncle Bert style'. Thus, much of the noise outside, which later developed, was obliterated. New Brighton was the chief target area again, and some heavy stuff was dropped quite close.

The next morning, Sunday, on the way to St. James the nearest damage was only 2 and a half-minute walk from Ormiston some 250 yards away. It was in a stick of 8 bombs; the two biggest fell in Grosvenor Road, demolishing a considerable area of property, and in the next road to it (a distance of 10 minutes' walk from my base).

I enjoyed the Whitsun at Church, although the choir and congregation were even more sparse than usual – probably as a result of the previous night's 'work'. Being as the type of service at St. James is low church, there are never any processions, and at all times there are no candles on the altar.

Again, that night the warning sounded, only 10 before 1 o'clock on the Monday morning. The alert lasted nearly 4 hours during which there was much gunfire, as on previous occasions, but no planes ventured directly overhead and kept closer to the Lancashire coast. Up to 2:30 p.m. on Whit Monday I was at the office, although a great deal of work was not done. Other days of the week, I do not see outside life until 6 o'clock or after. It is because of this pressure of business that my leave has been temporarily postponed, but any week may see me home after next one.

The question of razor blades has reached a point now when I must say that the 'Laurel' must be condemned and put away until such time as the Dumb-bell blade returns into production. The more common, or rather universal type of blades are the 3-hole blades in which there are still ample selections from which I choose. It will necessitate the purchase of a new razor outfit.

The landlady returned from Nantwich last afternoon after having a quiet time with only one warning and I am now back residing as a full resident of No. 3.

Regarding Clothes rationing, which came into force on 1st June, I suggest that you cut out the scale of ration coupons from the newspaper, and keep them in the old ration book by the side of the 'margarine' Coupons, for reference purposes.

I cannot think of anything further now, so I will shut up this epistle until I have collected further material upon which to relate, comment or discuss.

From your affectionate son,

[signature]

* * *

Letter 46

3, Ormiston Road,
New Brighton,
Cheshire
24th June, 1941

My Dear Mum,

At the station entrance, I was instructed to proceed to platform 3 for the Basingstoke train. Other officials of the Southern Railway later decided otherwise. Eventually, I finished up at No. 1 platform where I had to wait until 8:30 p.m. the Reading train was waiting at Basingstoke for the arrival of this London train, therefore there was no waiting there. At Reading I sat on a platform seat for the three-hour in an endeavour to keep cool. I soon fell asleep, and awoke on the noise of the arrival of my 1:10 a.m. train for Birkenhead. It roared into the station, 10 minutes before time. It was ever fuller than usual, and there was no seating accommodation. I sprawled out in the corridor, where it was cooler, and once more fell asleep. By 3 o'clock, I found room in a compartment crammed with service men. They turned out at Gosford at 4:30 a.m., and I then made full use of the length of the compartment, and laid out straight along the seat. I awoke at about 8:00 a.m. just as the train was entering Chester. For some reason, the train always wakes its passengers at this point.

I arrived at Woodside Station, at 8:45 a.m. where I had my breakfast before proceeding across the river to Liverpool. I reached the office about an hour afterwards.

The weather is overcast, and is considerably cooler than the South. I have not yet had time to consult anyone regarding the past week's weather conditions, or even aerial activity. Piles of papers awaited filing this morning, and in addition, I have to cope with showers of correspondence.

The tempo of work in the office does not appear to have relaxed during my absence, and there is still a possibility of abundant activity in this direction for some considerable period.

I have little more to say, and still less time in which o say it, but I hope you were able to get home before dark last night.

From your affectionate son,

***

Letter 47

3, Ormiston Road,
New Brighton,
Cheshire
4th July, 1941

Dear Mum,

Thanking you very much for the card, showing Winchester Cathedral posted in the same town on Tuesday, and received by me by the first posted on Thursday morning, 3rd July. It I did remind me of my glorious holiday and of the day (or rather part of a day) spent in Winchester. I think of that holiday daily, and I think a great deal of time will elapse before its memories will be eradicated.

When I returned to the digs yesterday after another day at the office (slightly less hectic than usual) at 6 o'clock, I was going to take another look at your card before tea, when I saw your schedule, weekly letter waiting for me – usually receive it Thursdays. The last paragraph took me very much by surprise and it was some time before I could recover my equilibrium. I have a rigorous habit of underlining important or urgent sentences with lead pencil, never before have I found it necessary to underscore with red pencil. I am glad this weekly letter did arrive so early as it gave me time to prepare an overnight for early dispatch on Friday morning. It was a great pity, in view of the circumstances, that my holiday did not fall due this week. However, I am enclosing a letter for Vera which is not as long as my usual epistles, the reason being that space of time does not allow for such.

After a spell of three weeks I attended my first choir practice last Friday, (27th) as I promised in my last letter. I went to the underground hall, where we usually assemble, but found the door locked. For a moment I thought that no practice was forthcoming. I crossed the road to the Church when I saw another member of the choir standing outside. He had told me that the week before, the practice was held for the first time in the church. However, as we stood outside I saw the little form of Mr. Smith in the distance coming toward the church. The practice was held in the Church, and I enjoyed it, when other members arrived. There were four men and three women when they ceased coming in. We are practicing a fine Anthem – 'who is like unto Thee, O Lord' by West. This week's practice is intended to be its final rehearsal.

In a scotch mist yesterday, I made my way to Liverpool Cathedral. This is the first time I have entered this great building since I attended the short organ recital on Easter Monday afternoon. The last time I came I walked through the neighbouring roads to enumerate the damage there. It is interesting to note the determination which is behind the Cathedral builders; in fact, it is a wonderful display of courage. Despite the loss of sheds, cranes, and other machinery necessary in the constructions of such a fine edifice, and the damage done by bombs, yesterday I saw that sheds had been replaced and cranes are again swinging masonry into place. In fact, there are many signs of great progress during the past two months since I last saw the place. The great tower of the Cathedral which can be seen for many miles on a clear day, has grown several more feet skyward, and it is now being crowned with ornamental battlements, at a height of 440 feet.

The large window in the South Transcript was blasted in May, and now, a scaffolding stands outside the empties window. Inside it appears that another great event in the Cathedral's building history is to be made shortly. Deconstruction of the screen separating the recent part of the Nave, from the choir has already begun, enabling the two parts to be joined in one. Quite a quarter of this screen and supporting staging has been removed.

I have commenced seeking fresh digs. The landlady has hinted at this several times since my return, in an endeavour to stir me into action. The crux of the matter is this. Her younger son, |Rodney is fretting in |Nantwich, and after much parlaying his mother has promised to bring him home. That promise was made several weeks ago. Since then, more recently as you probably know, a government appeal was made in this connection, requesting parents to leave their evacuated children and not to bring them back. Hence, to obey the promise, and to obey the Government put me in a precarious position. The solution is this. Mrs. Wills does not wish to live so far away from Nantwich but she is prepared to go to a nearer country area with Rodney. Therefore, for this to materialise, I must depart. When I have a new address I will advise you, although I expect it to be a week or so before I meet anything satisfactory.

Private correspondence this week has engaged a tremendous amount of my attention. Two cards were received from 'Francsland' and replies typed. Not being sure of the name of the lane, road, street, etc., I am enclosing the letters to yours. You will, no doubt, do the necessary. Correspondence with Great Yarmouth has opened up. It is with my old friend, Jim Hann. He wrote from the RAF station. I have replied with a four-page epistle. A letter from Parkstone has also been received and replied to. Now, I have your card and letter to deal with which requires an enclosure to Vera.

I am glad to say there is nothing to report in connection with raids, other than two warnings during the past week, since I dispatched your letter.

I will send my soiled linen, on Monday or Tuesday, after I have changed my second week's clothing. I trust you are keeping well under the favourable conditions, and that Vera enjoys her stay at Totton.

From Your Affectionate Son,

Edwin

***

Letter 48

3, Ormiston Road,
New Brighton,
Cheshire
8th July, 1941

My Dear Mum,

    I am enclosing my spoiled linen for the past two weeks. I now have one shirt and the necessary collars left. In addition to the shirt which I am wearing. Do you usually deal with the washing immediately on arrival or wait until Monday?

    The weather has been glorious, especially on Sunday, when New Brighton looked like a peace-time seaside resort in full swing for the season. The beach was crammed with trippers, and as I walked amongst them, I found it difficult to impress myself that I was not one of them: that I was a resident. That Sunday afternoon, the Windgate Temperance Band opened the first programme of the Band Season in Vale Park. There was a good programme, and I browned myself off in the sun with Mr. Moody, with whom I had made an appointment on the previous day.

    Correspondence continues to blitz me. On an average, during the past week and up to date, I have had two letters a day. Letters between Great Yarmouth have commenced in earnest, and according to a suggestion by Jim, we shall establish a weekly system.

    Possibly, by the time you receive this Vera will have returned, but I wonder how she is enjoying herself at our noble residence and how and what does she think of the ruins of 'the town on the South Coast', or 'a town in Southern England'. By the way, this morning, over the radio I heard that 5 aircraft were shot down during a heavy raid on a town 'in Southern England' last night. I trust this does not mean a repeat of that Saturday's 'do' while I was there.

I hope you have been able to dispatch the letter to 'Francsland' – I have had another from that address, (Francsland, Hammonds Green, Totton). It seems to be very vague to me, as Hammonds Green is a district of Totton, like Millbrook is of Southampton. Does the postman understand the address?

Yesterday, a warning sounded early in the afternoon. I slipped out of the office to post an epistle to Jim Hann, and when I dropped the letter in the Post Office Box, I heard it drop, as I thought, with an extraordinary bump. I walked away, and heard a similar bump. I soon discovered there was a raid on. The bumps I had heard was gunfire, and these were followed with several others. There was a plane overhead, but it was soon disposed of and the alert only lasted 30 minutes.

From your affectionate son,

***

Letter 45

3, Ormiston Road,
New Brighton,
Cheshire
11th July, 1941

My Dear Mum,

I thank you very much for your welcome, interesting, and long letter. By now then, Vera will have arrived, and I shall expect to receive some communication, but until then I shall remain mute.

I have been anxiously awaiting the results of the raid, and your letter has now satisfied me on that point. The Millbrook Recreation Ground seems to act as a magnet to high explosives.

***

Letter 46

> 3, Ormiston Road,
> New Brighton,
> Cheshire
> 18th July, 1941

Dear Mum,

Thanking you very much for the parcel, which I received on my arrival from the office last evening. The linen has arrived in comfortable time for Sunday. I also extend my thanks for the chocolate, and the Southern Daily Echo, giving a description of your last great raid. I also noted in that paper, how well advertised was the Registration date for the nineteens.

I was pleased to receive some news of V.H., after such a considerable time since the first intimation I received of her intentions at the beginning of the month. I have not yet received the letter promised in your parcel-letter, but no doubt something is in the post by now. I see you are acting on my suggestion of obtaining a photograph, which will I hope turn out to be a good one. It is too late now for me to offer any suggestions regarding suitable settings for the photograph.

Events of the past week have turned out to be varied and extreme. My history, continuing from the dispatch of your letter last Friday went on as follows.

On Friday night I attended another choir practice which up to the time of commencement did little to give me much optimism regarding the attendance. However, the attendance was super, showing an increase on the previous week. There were 5 men and 3 ladies. After dealing with Sunday's hymn and chants, we proceeded with an Anthem by Coleridge-Taylor, 'O, ye that love the Lord'. After squaring up many points there, we concluded with West's Anthem, which should have been performed the Sunday before last.

On Saturday afternoon I paid a visit to the local employment exchange at 2:00 p.m. Around that building, lay the remains of many squalid slum dwellings, and the base of a Roman Catholic Church. It is not far from the Exchange Station, which is, considerably damaged, and the Bradford Hotel. Inside was a counter with three clerks behind, confronted at some distance by two rows of a dozen chairs each, filled with youngsters (positively of my own age!!!) but not a very high intellectual capacity, which explained their apparent bewilderment in those surroundings. A policeman was in attendance and on my entrance, he conducted me to a 'pew'. We were dealt with in order of arrival. Within 20 minutes my turn came to go up to the counter to be questioned by one of the clerks behind it. I supplied the necessary answers, and explained that the address on my identity card was not my home address. They explained back by saying that the address in New Brighton was the one they would take, as it would be the best for communicating with me, at a moment's notice. From the counter I was directed to an annex where my occupation was categorized – as shipping clerk. That is the history of my registration for the Armed Forces, July 12th, 1941 – 'With a reference for the Royal Air Force – Ground Staff'.

That night it thundered and lightening from 10 p.m. until 5 a.m. the next morning. But I can say quite truthfully that I slept through it all, except the first few minutes when I was going to sleep, and at 5 o'clock, when the force of its final clap of thunder woke me up.

On Sunday afternoon, in the sultry heart, I listened to the Brighouse and Rastrick Band in Vale Park, which rendered a good programme to many people. The programme included Ketelby's 'in a Persian market'.

Sunday evening at church for Evensong, dark clouds gathered around, until no light came through the windows at all, only the electric lights shone outwards – as if someone had omitted to put up the blackout, at night. Towards the end of the sermon distant rumbling could be heard, and the brilliance of the lightening increased as the storm drew nearer. Later during the course of a hymn, when that 39 stopped organ, 3 men, 3 women and 2 boys were going full blast, to say nothing of the congregation, the peals of thunder crowned all! Deluges of rain fell and found weak spots in the church's blasted roof, and shattered windows. We had to wait sometime after the service before the storm abated, to enable us to go home.

Yesterday I walked out of the office with a shrewd little packet under my arm, and another in my pocket. No one could guess what they contained. The external packet contained half a pound of BUTTER, presented to members of department each.

The other was a RAZOR which I obtained from one of our 'bubs'. The landlady was surprised to receive the butter, which I said would augment our butter ration considerably.

On the Sunday of my birthday I obtained the People mainly for the purpose of a general forecast of my nineteenth, or rather 20th year of life. The main point was a prediction of a decline in my financial position, and that my income would be considerably below par.

That at first was more or less obvious to me. As, by my being taken into the forces, I would lose the subsistence hitherto supporting me in Liverpool, and my only income would be just the bare salary, which even with my first rise, is not anything much to write home about. But, another unexpected turn in the finances is to take place. On September 1st of this year, the subsistence is to have a drastic cut very nearly 50%, i.e. a 25/- per week reduction, leaving only 30/-. That is for unmarried men. The reduction for married men is by no means so great, and what is more their salaries are 3 or 4 times greater that juniors. If I were married, with a wife in Southampton, I would get more, financially, and leave more often (once every 4 weeks). However, that comes into operation in September, 6 weeks from now. The dig situation will receive considerable attention from me during this forthcoming weekend, should the weather favourably permit.

Work here at the office continues to be a blitz. We have had two more arrivals during the week, which has not done anything to relax pressure. But I must say, that now, I have various systems working in the various branches of my work which see me through the most troublous times. In fact my Chief Clerk, frequently says 'You're coming on, boy'. The last word is always stressed. I still find my business day extends itself to 6 o'clock and after each night, and I do not arrive home much before 7 in the evening, then I have to have tea, before I am free to entertain myself.

Although I knew while I was home, the date on which I should carry out a fire-watch, I refrained from mentioning it, until the duty was carried out. Last Tuesday night was the night. Mr. Hume also was included with the fire watch. The system is slightly different from India Building. The main difference is that there are an equal number of bank staff and tenants on watch at the same time. The squads are dividing into twos, each two containing one bank man, who takes all the responsibility. We have had a lull of 10 days without a warning from last Wednesday week. The continuance of rainy weather may, even when Jerry does decide to come this way, have a deterring effect.

It is five weeks since I last came down on leave, and perhaps within another 2 or 3 weeks, I shall be down again for another week (10 days).

That is the end of the news. Hoping you will keep well and safe.

From your affectionate Son

***

Letter 47

3, Ormiston Road,
New Brighton,
Cheshire
29th July, 1941

My Dear Mum,

Thanking you very much for your interesting letter which I received this morning enclosing one pair of socks.

On Monday of this week I received a letter from Auntie Ruth saying that she would be having her holiday, commencing from Thursday 31st July – the same day as mine commences, or at least it is my travelling day.

I am experiencing great difficulty in seeking accommodation in New Brighton to last me for the 3 weeks when I return, while I play at St. James. If I cannot fix up I shall have to cancel the appointment for Mr. Smith. I am due to see him tonight at 7pm, and if I cannot get any satisfaction before that hour, the appointment will be off.

The majority of people do not wish to take the responsibility of housing boarders in these times of Blitzkriegen, and food rationing. Many refuse on the grounds that they are leaving the district.

Last evening, I went to a cinema show at the Trocadero in Victoria Street, New Brighton. The main film was 'Philadelphia Story'. The producer's name ends with '...ez', but not being an habitual cinema goer, I have forgotten the name. I did not go alone. The chief star was Kathleen Hepburn. Two scenes in my opinion were unduly protracted.

Last Sunday, the weather was gloriously fine. In the afternoon I sat in Vale Park and listened to a Military Band. On Sunday evening we sang another anthem – 'O, ye that love the Lord' – Coleridge-Taylor.

Providing then, there be no sudden cancelment to my leave, mainly owing to unforeseen 'arrivals' I hope to be seeing you either very early Friday morning, or by mid-day.

In that case, this will be my final letter of my 9th term in the North West. From your affectionate son,

\[signature\]

***

Letter 48

22, Park View
Waterloo
Liverpool 22

Dear Mum,

Once more I have arrived in the North-West. The journey was uneventful. I arrived at Birkenhead at

8.30, which is the earliest time I have arrived from my leave. I suppose you arrived home before it was dark,

I will be busy at the office for a day or so until some outstanding items, which I only can deal with, are squared up. There have been no raids during my absence, and it is only hoped that the peace will continue.

Tonight I shall install myself in my new digs, about which I cannot say any more at the present. This letter commences my 10th term in Liverpool.

From Your affectionate Son,

Letter 49

22, Park View
Waterloo
Liverpool 22
15th August, 1941
(Friday)

My Dear Mum,

Having finished my first day's work at the office, I was escorted by Mr. Carpenter to the Bus stage. The bus stage is by the Lime Street Station, where, while sitting on the top deck of the bus, looking into the station, I saw with envious eyes the 5.25 train depart for Euston, after a good thirty-minute bus ride through the districts of Liverpool enumerated to you while I was on leave; districts which revealed many results of modern warfare, I arrived in Waterloo with my case. This district is truly on the outskirts of the City. It is served by the Ribble Bus Services and also by an electric Railway. The latter route should reduce the travelling time by half.

The present 'embassy' is a large house in a row of dwellings overlooking a Park. Its inmates are the landlady and her daughter; other residents being, a young married couple, the husband being in the Merchant Navy, serving in a Line which held a great deal of interest before the war. My room-mate will be the musically minded and also brain-minded fellow, who I mentioned while at home, who is away at present. I have another room mate at present, who apparently lived a good deal of his life around this district. There are two other gentlemen, both keen draught players. The victualing is on the same standard as that at my former residence in New Brighton.

Before I had time to curl up in my new bed, for the first time, on Tuesday night, the alert sounded. It lasted 50 minutes, but no aircraft passed in the vicinity. Shortly after the 'All Clear', there was another short alert.

On the 13th, the next day, I awoke to find the world outside a wet and windy place. At 7:30 a.m., I commenced Breakfast and finished the repast by 8:00 a.m. I set out into the weather and walked to the bus stop by St. Faith's Church (a seven-minute walk) and there I became sufficiently damp to set my mind running on thoughts of 'the inconveniences of Liverpool transport', as bus after bus rushed past in a cloud of spray. At the end of 30 minutes I felt sufficiently well damped, that I leapt at the next bus that appeared, heeding not whether it was full or not. However, I did arrive at the office that morning – after 9 o'clock! Of my own accord, I quickly defeated the riddle of the Ribble Buses.

That evening my room-mate conducted me on a tour around the surrounding roads. There was a gale blowing from the sea which made the expedition none too pleasant – for my lid, which on two occasions yielded to the will of the wind, rather than to the will of its owner.

I received my first letter yesterday morning, Thursday, at this address; it was from Auntie Ruth. In the evening I received your first letter of my 10th term for which I thank you. I see it was posted at 1:00 p.m. on Wednesday.

The subsistence question was reviewed yesterday by our MANAGER, who called together the heads of all departments to discuss the positions of the juniors of the staff. He considered that in many cases considerable hardship may be entailed by them when the time comes from the deduction in subsistence to be affected. It must be born in mind that the reduction is forty-five percent of our wages. Mr. Child of our department therefore called C.A.R. and myself later to his room, and explained the situation to us, and said if we put a statement of our expenses in to him he was sure that Mr. Yarrow would approve. We then made out the statements showing the total of weekly salary and future subsistence, less every day expenses, which left a very small remainder to cover the cost of clothing, toilet requisites and other items. Our two cases have been accepted as fair and just, and will be recommended.

I visited Liverpool Cathedral on Wednesday to find it in its new and altered form. The original entrance door is now closed, but one on the opposite side of the South Front has been opened for the admission of the congregation. This leads to the newer portion and from the west end of this for the first time I have seen the entire length of the interior of the Cathedral in one glance. It would be well worth you seeing when you come this way. As the building grows so it becomes more and more impressive in its massiveness of style. It is, incidentally, just a year ago that I first entered the building. Externally the battlements of the tower are still growing and the two dwarfed looking cranes on the summit continue to hoist their loads of masonry into position, with minute men crawling about among the scaffolding. The whole place seems to be alive with industry.

In the evening of the same day, a colleague kindly called at my lodgings in Ormiston Road, New Brighton and reported to me the following day that there had been no correspondence for me since my departure; hence Herr Hann has not yet settled, and yet V.H. has not written – this is 6 weeks since I wrote my last letter to her.

I hope you enjoyed the Organ recital at the Southampton Guildhall last Thursday, and I hope to have your reports on these weekly recitals, in the same style perhaps, as those I wrote regarding those I attended at the St. Georges, last autumn.

Apparently, Mrs. D [landlady] has told someone at No. 22 that I am partial to music, as on several occasions during the week I have been requested to cause vibrations in the air through the medium of the Moore and Moore piano. I am quite uncertain why I have been so stubborn in response to their appeals. I just won't play – yet!!

Each day, my mind reflects the events of my past leave and when I am not thinking of those glorious moments my mind turns to the events of the past year when I was all eyes and ears in a strange world, among strange people, with strange habits and customs. Here I am today, still in that world of dirt and squalidness with its massive imitations of Greek styles in architecture. I have seen during that year, a steady change in the landscape as a result of modern warfare, until today when whole areas are laid waste in piles of debris. This time last year I well remember our instability in India Buildings as an office staff member. We had not yet received all the necessary office equipment. By now, I had spent nearly a week at the Bradford

Hotel, and was soon to move to No. 1 Manville Road, where I was to join Mr. Moody. However, here I am.

So, I will leave you, and close my first weekend letter of my 10th term in Liverpool, hoping to hear from you again next week.

Oh, there is one more important point. I have welcomed the great news regarding the activities of the Air Force over the enemy country and enemy occupied territory, I expect you heard some aerial activity on Wednesday.

From your affectionate Son,

***

Letter 50

22, Park View
Waterloo
Liverpool 22
15th August, 1941

My Dear Mum,

I thank you once again for another weekly letter. There have been few events at this and since you left on Monday. The only aerial activity during those 5 days having been an alert on Tuesday afternoon, which many people took to be a prelude to some night's activity. I know of one person who left Wallasey for the night and installed himself in Bebington, but returned the next morning, as nothing materialized.

I have carried out two more reconnaissance visits to Christ Church in Waterloo Road. On both occasions I examined as much as I could of the instrument without forcing the lock! I enclose an uncompleted specification of the Organ, for your records.

On Monday the Victualing Dept., are to have a new typist, which no doubt will be another pain in the neck to me.

Draughts have been indulged in and operations proceeded according to plan. I have also spent a few hours at the piano at the above address, on occasions when the house has been empty in the evenings.

I am so glad you had such a quick and easy journey home. At Euston you need not have left the station, as you could have descended a stairway to the Underground Railway which would have conveyed you to Waterloo Station within 15 minutes. Did you see the dirty approach to Lime Street station as you left. I am enclosing an interesting resume of the visit to Liverpool. Where I have written the word 'you' substitute the word 'I' when reading it to dad. Read it as it is written to yourself first. You will be surprised at what you achieved during the two days.

The weather is wettish and windy. I close now, until next week. I forward my laundry under separate cover, on Wednesday, not enclosing a letter as I am anxious not to delay it longer.

From you affectionate Son

[signature]

[This was attached to the letter]

The following is a resume, in a succession of statements of your visit from Saturday 23rd August to Monday 25th August, 1941. (when reading it to dad substitute for the word 'you', the word 'I'.)

At last you have seen the city of Liverpool. You have seen its high-ways, its narrow streets, and cobbled by-ways, full of hurrying, queuing crowds, boarding tram-cars, packed in buses, or cruising by in private cars.

You have seen the great commercial buildings, encased in many layers of soot and grime, in which the business of the city is carried on. You have seen the closely packed dwellings and tenements in the much congested and devastated areas of the city, Bootle and Wallasey. You have seen many of the ecclesiastical buildings. You have seen the St. Georges Hall, Martin's Bank Building, India Building, The Royal Liver Building, Cunard Buildings, The Mersey Dock Board Offices, and the Adelphi Hotel, Lewis's (gutted) stores, Blackler's (gutted) stores, Bon Marche and Hughes.

Above all, you have seen the great Cathedral, and attended a Sunday morning service. You have seen the choir boys and men in the coffee coloured surplices and red cassocks; and the servers in green robes, within the surroundings of immense columns and archways, built of the russet-brown stone quarried near Childwall. You have heard the largest Cathedral Organ in Great Britain, played by Mr. Goss-Custard.

* * *

Letter 51

22, Park View
Waterloo
Liverpool 22
1st September, 1941

Dear Mum,
I have duly received your parcel enclosing letter and chocolate (the latter does not now exist), for which I thank you. I believe the resume which I enclosed, fully carried out its purpose of recalling the facts to mind of your Liverpool visit.

The whole of this week has been eventful. Last Sunday morning I took a walk into Blundell sands, north of Waterloo, and on the way to Crosby. It was a glorious sunny morning, with a gentle breeze blowing from the sea. All the avenues in this charming district, possess trees on either side of the road, in the same way as the Avenue at home. The houses, too in these avenues are similar.

In the evening I attended the evening service at Christ Church, Waterloo, in Waterloo Road, near your lodging. The organist is a brilliant player, and displays more expression than the organist of St. James, New Brighton. The pedal department of the organ, is certainly very cathedral like, owing to its having so many stops allotted to it. I enjoyed the service, although I considered the texture of the singing of the choir, a little harsh.

Then, as the organist played the last Bach Prelude in the First Book of Bach, I boldly requested the Verger to introduce me to the Organist, after he had finished his concluding voluntary.

I was speedily conducted to Mr. Mason, and formally introduced, when I was left with him to say what I would. Within five minutes of meeting him, I was singing in a choir practice, which apparently customary after the evening service. Mr. Mason is a very likable young man, fair curly hair, and blue eyes. He is the son of the old organist who has recently retired. The old gentleman is now in the choir, and also likable. He is a dear old English gentleman, but owing to his age, and a recent accident is very tottery.

There are 20 boys in the choir supported by 12 men! – Just 31 more members than the total of 1 in Millbrook) I touched on the subject of Organ Practice, and Mr. Mason thinks that it may be possible for me to have one practice occasionally. He said it is difficult for me to use the instrument at the moment owing to instability of the tower of the church. In that case thank goodness, the Trombone 16ft is not installed – or there would be no church left if that stop came into operation with the others!

At the practice e, an anthem was sung, you know it, Stuart knows it and I know it. It reminds us all of the good old days – 'how lovely are thy Messengers' – Mendelssohn

As from Monday, the First of September in this year of war 1941, (1) the commencing date for the new rate of subsistence. (II) the arrival of a new typist to our Department, to whom I am tutor. (III) Joan, the landlady's daughter commenced work at a technical school on the Contometer.

In connection with:

I. *Wednesday was a notable day. I was requested by the Manager, Mr. F.C.V. Yarrow to visit his office. This was the first occasion I have been called to his room. (now start thinking that I am going to get the sack!) Mr. Yarrow questioned me from various angles regarding the subsistence question, and concluded the catechism of consenting to recommend to the Managers in London for a further consideration of the matter. Actually, everyone was called to his office, individually, this week, those who had made statements regarding the subsistence, and there were many.*

II. *is the bane of my life, and thank goodness I have not the prospect of the process continuing very long; that is the days of my office life are numbered*

>   III.   is my morning escort on the L3 or L1 bus from Waterloo into town.

Al these three points commenced on September the First, 1941.

On the third of September 1941 is the second anniversary of the outbreak of war, I commemorated this even in orthographical form, writing an epistle to Stuart on 9 pages of this paper. It was an answer to his letter of some 10 days ago.

On Thursday morning I received a letter OHMS notifying me of the date and time and place at which I am to submit myself for medical examination. This will be my third medical examination in three consecutive years. The exam is to be on Monday 8th September, at noon, at Renshaw Hall, Renshaw Street, Liverpool. The next piece of correspondence which I shall expect from the Government will be the papers informing the town place, and the time to which I shall have to report for Service. I am looking forward to the possibility of change in occupation and scenery. Many people so often take a pessimistic view of this negotiation but to me, I feel, that it will be a continuation of my education and in a further broadening of my outlook, which has been set afoot in Liverpool. I hope to see many other parts of this country of ours as a result, when I am called upon to render my services in the forces.

Although that time may not be very far distant, I hope that it will be possible for me to obtain a weekend leave before then it is now 6 weeks since I was last granted leave by the Company, and really to complete the full complement of weeks there would be two more to go. In any case, whether I shall come home next week or the week after, I hope to travel down on Friday, arriving late night or early Saturday morning, which will make it a Monday for me to travel back again it is more possible for my leave to be a week later, i.e. 19th September.

Well goodbye until next week, when I shall be able to tell you more about the medical.

From your affectionate son,

[signature]

***

Letter 52

>22, Park View
>Waterloo
>Liverpool 22
>10th Sept, 1941

My Dear Mum,

I am enclosing my soiled linen which has collected during the past fortnight.

I have at last sat in the choir stalls at Christ Church, Waterloo. It was the National Day of Prayer, and the choir recessed at the end of both the morning and Evening service. The total number of member in the choir amounted to 35. There are 20 boys, 12 men, and 6 women – What a choir! It was not unlike a small Philharmonic Choir. I am a tenor on the Cantori side (organ side of the Chancel). In the evening the choir sang an Anthem – 'Lord, maker of Heaven and Earth'. In which there was a Tenor solo, which was taken by a Tenor on the Decani side. The anthem was an arrangement for men only.

On Monday, I attended the Medical Examination as requested by the Medical Board at mid-day at Renshaw Street. After submitting myself to five doctors in succession during the greater part of three hours. I then interviewed by Air Force and Military Officers, respectively. The whole examination took 4 hours.

The typist is still at the office and is still the bane of my life. As I have already mentioned it is my misfortune to be her tutor.

Today, for the first time since my arrival in Liverpool, I have visited one of our vessels. I thoroughly enjoyed viewing the ship as I used to years ago when at home. When I come home on leave I may be able to say more about it.

I hope to come home next week as it is the 8th week since I last travelled down, which is the full complement of time. Providing the Company grant the weekend leave I should be travelling on Friday, the 19th September.

I received a letter from Auntie Ruth, and a letter from the organist Association on Monday. I replied to the former the next day. I will close now, until my weekly letter.

From Your Affectionate Son,

[signature]

***

Letter 53

22, Park View
Waterloo
Liverpool 22
12th September, 1941

Dear Mum,
   I thank you for your letter which I received on Thursday evening on my return from the Office. I have not heard anything regarding the Subsistence question.
   There is really little more important news to impart to you, other than to mention that I have encountered several fiercer battles on the draught Board since I wrote my last letter on Wednesday, 10th inst. Both my offensive and defensive strokes have told heavily upon the enemy in an overwhelming majority of battles, resulting in many stunning victories.
   Last night, I had a mathematical fit. The whim struck upon Geometric Progression in an Algebra book I found in a bookcase in the drawing room. I, eventually involved the school master in my calculations. I clearly remember one lesson at College when Mr. Elsmore demonstrated before our Senior Commercial Section, the number of possible positions 4 people could sit in a carriage compartment with accommodation for 8 persons. With the draught board out on the table I thought of this query. If on the first square I were to place a half-penny, on the second twice that amount and on the third square twice the last amount, now much money would there be so on the last square? It was greater than you would suspect at first glance.

The contometer is a machine included in Office Equipment, for deducing mathematical calculations. In appearance it rather resembles a typewriter, instead of letters on the keys there are numbers, and various levers for effecting addition, subtraction, multiplication, and division. It is operated similarly to a typewriter. In my book case there is a book with an orange coloured cover in which you will find photographs of the contometer, together with a fuller description than I have given here.

When the time comes for my final departure from the office, and Miss X takes over my duties, I dread to visualize the consequences. I can see the corruption of many of my infallible systems, resulting in such abundance of chaos that the patience of the seniors of the department will be taxed beyond normal strain, and will result further in violent action, and no doubt, in addition, a volley of words which will poison the ears of their recipient.

I am whole heartedly sorry for the future of the office as regards to efficiency in the junior section.

Love,

[signature]

* * *

Letter 54

22, Park View
Waterloo
Liverpool 22
15th September, 1941

Dear Mum,

I thank you for the parcel enclosing a clean shirt for next week, which I received last night, Tuesday, 16th inst. It only took one day to reach me.

I visited our General Department at the office yesterday in respect of my weekend lease, but according to their calculation

I am not due until the week after that (26th September, 1941). Next week I will again discuss the matter with them. It is perhaps as well that I am unable to obtain permission to come home this weekend, as I have a slight cold and excessive travelling as is necessitated in coming home, would not be advisable under such conditions. I have resorted to the necessary quack reliefs, which I have had by me in my case. I have used the Vick for the first time since my evacuation to Liverpool. Colds are quite prevalent just now.

Last Friday the much-anticipated return of Mr. R. H. Cranmer-Gordon, familiarly known as Booby, was materialized in his appearance that afternoon. It had been felt generally during the time of his absence that he and I would be well suited to one another. He is short, close on 18 years, and displays much self-confidence. At present, he is being educated at the Merchant Taylor's School, where he is studying a higher course, including French and German. He is now my room-mate, the schoolmaster being transferred to another room. That room now hears the lengthiest of discussions, and debates on musical, psychological, and optical subjects.

Shortly after his arrival, he indulged in music, the chief item being the Konzert of Grieg's, a brilliant but difficult work proceeding to great length. It is written for the piano and orchestra, and this week we both sat at the piano, he taking the piano parts and I the easily sight-readable orchestration. There is no doubt that the composition is far above the heads of all the other members of the establishment, but they had to grin and bear it. In fact, that was last night. Bobby played two of his own compositions a Prelude and the makings of a Concerto. He then went on to demonstrate works of Rachmaninoff, including Concerto No.1 and Prelude No.2. I followed up with a memorization of the C Sharp Minor.

When residing at No. 3, Ormiston Road Mr. Moody found that whatever I played the C Sharp Minor Prelude, it was always followed by an alert. What would you expect to happen after so many works of that composer had been dabbled with in one evening, by two people, at the same house!! Well, the warning did go, and it is the first time since you were here.

During the earlier part of last Sunday, I registered for Fire-watching duties, incompliance with a request by the government to all Citizens in target areas. I was given an exemption form to fill in but the lingo used was beyond me, and hence I am forgetting the subject. In any case by the time, the scheme is brought into operation. I will doubtless be moved from this area, by another part of the Government.

Owing to that registration, last Sunday morning, I was unable to attend Matins at Christ Church, but I sang in the choir in the evening.

Draughts have been substituted for Whist. In the former occupation, there are two opponents needed, in whilst there are two or more. On many occasions, I have had outstanding scores. On Saturday, my partner and I had a lead over the other opponents but a quick change in our fortune allowed them to catch up to us, and later we had a further spurt and reached 20 long before them. Last night, I rained ace after ace, in the earlier part of the game, reaching a high score before our opponents had scored any tricks. Later our rate slowed down and they just passed us reaching 20 first.

You will not now have to expect me until FRIDAY NIGHT or EARLY SATURDAY MORNING, 26/27th September. I should like to have a crack at some of the piano pieces I have, and that are at home. Especially the Introduction and Fugato: the rustle of spring, etc. you need not send these on, but when I come home I will change them for some of the remaining music I have here, which is not much.

From your affectionate son,

***

Letter 55

22, Park View
Waterloo
Liverpool 22
3rd October, 1941

My Dear Mum,

The opening of the 11th term in Liverpool marked the beginning of a recuperation period after the short sickness at home, during my 10th weekend leave.

The return journey via Reading was uneventful, although the Birkenhead train did not arrive as early as some of its predecessors. On Birkenhead station, *I had my breakfast, comprising of a good deal of victuals supplied by you in the case. A cup of tea was relished. Already comparatively late in starting the repast, I made no effort to catch up time, but concentrated more on having a good meal. I arrived at the office in Liverpool at 10 o'clock. Many discrepancies faced me as the result of insufficient efforts on the part of the newcomer to continue the normal office routine during my absence. The staff were pleased to welcome my return, to right the wrong and resurrect the office system.*

That night I returned to the digs and after having dinner, retired to bed shortly before 8 in the evening. Before 10 o'clock there was a short alert which disturbed me, otherwise I had a good sleep until 7:15 a.m., the next morning.

This day was your birthday anniversary day, and I celebrated it here in the North-West. Firstly, by accidently biting my tongue, in an already weak spot at breakfast, causing the remainder of the meal to be abandoned. At 1 o'clock, I dined at a café which I had not visited before. It was Reeces over the other side of Water Street. It was a fine, hot lunch comprising Soup; Cottage Pie, sauté potatoes, cabbage; baked jam roll and custard. It was worth 1/9d.

At the digs at 6:30 a.m., I was enjoying a dinner consisting of 4 sausages, pots and peas, apple tart and custard, and tea. After that consumption of food, I considered a walk in the open air favourable. I, then returned to my base and prepared this letter.

However, I hope you will be able to obtain the statuettes which will be my present to you for your birthday, the second since my absence from home.

This is the first letter of my 11th term.

From your affectionate son,

***

Letter 56

22, Park View
Waterloo
Liverpool 22
14th October, 1941

Dear Mum,

I am enclosing my usual fortnight's laundry.

At Christchurch, Waterloo, last Sunday, the presentation to the retiring organist, Mr. Mason was made. Members of the choir and congregation who wished to attend the presentation assembled in the choir vestry after the morning service. The vicar called upon Mr. Holland, the present tenor soloist to make the presentation. Mr. Holland made a short but eloquent speech. He said he had been in the choir for 40 years (since 1901) and could remember Mr. Mason as organist, even when he was small singing in the choir. He complimented Mr. mason on having good qualities, not only as a musician, but also that he possessed a fine personality which never failed to hold those who served him. Mr. Holland then presented an envelope to him containing a cheque which he said was only a small token of the gratitude they wished to express.

Mr. Mason thanked them warmly saying that their great gratitude touched his heart. He gave a brief summary of his 42 years he spent amongst the choir at Christ Church, and he was proud of the way in which his men had supported him in making the choir, second to none in the district. Boys had come and gone, but the men still held fast. Mr. Mason then spoke of his blessings. He said he had been blessed all through his life with good health, and a love of music. He was glad his son had taken over. He, then thanked everyone for the very kind gift to him. There was loud applause after which the vicar thanked those who had attended the ceremony.

In the evening of the same day, there were two organists in choir. A young man now serving in the Navy, happened to be in port and came in to the choir. He has done so before when he happens to be in the district. He was an organist in Bath before he was called up. Our meeting was mutually pleasant.

Last Sunday night, we had an alert which lasted 2 hours during which a small number of aircraft passed in the vicinity. One unexploded bomb was dropped not far away.

On Saturday evening and Monday evening of this week I spent a fair amount of time at the piano and enjoyed playing the studies I brought back with me. There are prospects of our being busy at the office shortly. In connection with the office, the chief clerk has complemented me on my efforts during the past 4 months or so. He hopes the day for my departure will be far off as he feels that without my assistance he will have great difficulty to keep things running smoothly. He considers I am an important factor in the makeup of the department.

I have little more left to say now for my next weekly letter, so I will close down until then.

From your affectionate son,

P.S. I am enclosing one tie for exchange for one spotted blue tie.

* * *

Letter 57

22, Park View
Waterloo
Liverpool 22
24th October, 1941

Dear Mum,

I thank you for the parcel which I received yesterday evening, enclosing laundry, and an orange. I am reserving the latter for a few days. As there is little to reply to in your letter, I will get straight get straight ahead with this week's news.

Last Sunday I enjoyed a fair amount of choir singing at Christ Church. After the practice following the afternoon service I obtained permission to practice on the Organ for the following Sunday afternoon which will have materialized before you receive this letter.

The dog, after causing much havoc among the clothing of many inhabitants at No. 22, has died during the week. Personally, I am thankful as it was a great nuisance; eternally flying for one's trousers and tugging at them. It had had several screaming fits, which caused me to doubt its mentality on many occasions. The veterinary surgeon who took the animal away after death examined him and on Wednesday returned to give his report. He pronounced to Mrs. D that the cause of death was a result of the condition of the dog's stomach, which was very inflamed, also due to a wastage of his liver.

Air raids once more have entered the news. On Sunday 19th inst, there was a 10-minute alert, from 8:00–5:00 p.m. but there was nothing to report. The following night I was left in the house alone, while other occupants had gone to a cinema. I was then free to practice the piano to the greatest possible degree for my future practice on the organ. I had been practicing vehemently for some time when I heard what I thought to be a knock at the front door. If found on opening it that the sky flickering with gun flashes. This was the first gunfire we have experienced in this area since July last. It was raining, and there was a great deal of cloud, and people were running down the road. I spent a little while in a shelter but not finding the raid to be of a serious nature I did not remain. The raid lasted three hours during which about a dozen planes were heard and a mine was dropped which killed about 30 people.

On Wednesday while the vet was discussing the cause of death of Rex to Mrs. D and her audience the sirens sounded again at 8:50 p.m. I had heard a number of planes flying round for some little time before the alert. The barrage was slightly heavier than that of the Monday night but even fewer planes were heard. More activity appeared to be taking place around New Brighton on the other side of the river. The industrial town on that side had a few bombs, and a few people lost their lives. Both raids were very slight with immense lulls between each aircraft which came over the areas.

Everyone in our department is continuing to be very busy. On two or three occasions this week, we have not left the office until very nearly six o'clock. I expect by the time you receive this letter Mr. Small will have arrived, although there is always a strong possibility of delay.

I have tried to obtain some more of the 'Gillette' slotted blades, but have so far been unsuccessful, my attacks not being very strong or persistent. I should be pleased if some time, you could try to get some more for me at home. I still have a sufficient supply to last me for several weeks yet, but I was thinking of the future.

I have commenced cutting another new tooth. This time it is on the bottom jaw, which will serve to meet the other new tooth I grew in March last. Both teeth must be wisdom teeth, to appear at my stage of life. The new tooth has only just broken through in one corner, and at the moment is no more than a pin point.

I am forwarding a Birthday card to Grandma in the same post.

From your affectionate son,

Edwin

***

Letter 58

22, Park View
Waterloo
Liverpool 22
10th October, 1941

My Dear Mum,

I thank you very much for the parcel which I was pleased to receive yesterday. I was surprised to find so many enclosures, in addition to the usual laundry items. I thank you very much for the chocolate, the 'Gillette' blades, the orange and the small bottle of Krushens Salts.

Your 'Birthday party' must have been a very sober one with no one there! However, I hope you will be able to find the statuettes you require, and that they may stand in future days as monuments of peace to your life, as their attitude usually suggest.

I enjoyed two brilliant services at Waterloo, last Sunday. It was the Harvest Festival of Christ Church. Considering the comparatively small area in which to produce vegetable crops in this district, it was surprising to see the church so well laden with apples, carrots, marrows and even oranges. With an abundant distribution of flowers, the church appeared more like a greengrocers shop than the background to a religious festival.

The morning service included an anthem 'Sing, O Heaven' by Arthur Sullivan, which I am sure was presented more artistically than the average work of the choir. Even the boys discarded their harsher tones for the occasion. Two hymns were sung with descants, with ample organ accompanied, and the result was very effective. Before the afternoon service there was half an hours Organ Recital by Mr. R. Mason. The service commenced at a rather inconvenient time of 4 o'clock, but nevertheless I was able to arrange for the tea at the digs to be ready on my return. Both at the morning and evening services there were recessional hymns.

In the middle of the week, His Majesty the King visited this port. From the point of view of sightseeing I was fortunate as the King was to visit the building adjacent to ours – Derby House. It was not difficult to find a good position in our building, or even to find an easy way to reach a coveted position. I found access to the roof above the Third Floor, through a window on the staircase. I walked round the root to the corner opposite Derby

House. I had a clear Bird's eye view from the show. The narrow side street passing the two buildings already mentioned was lined with a navel guard of honour, and a thin crowd on the pavements. This, the most important section of my picture, stretched away almost in front of me.

Two minutes before midday the crowd at the end of the street, where it joined Chapel Street, raised a cheer. Then a police car appeared around the corner, closely followed by a Rolls Royce (DTD 955) bearing the Royal standard over the radiator cowl. Several large automobiles followed. The Royal car was a saloon. It drew into the kerb and the King alighted at the entrance, which from my good position afforded an excellent close-up view. The King was dressed in Naval uniform. He bowed slightly to an admiral, and then disappeared through the doorway.

That night I carried out another fire watch, by throwing darts at a dartboard, and learning snooker on a billiard table, and indulging in table tennis.

Thursday was a very wet day. It was raining when I awoke from the fire watch, and it continued to rain hard until late evening; it was rain like we have in the South, whole-hearted rain.

Has Dad started smoking a pipe yet? I see Woolworths here have recommenced selling these articles this week, at 2/6 apiece.

On Tuesday of this week C.A.R. brought a letter across from New Brighton. It was from Mr. Torr at Hay. The letter was in reply to mine dated the 1st of April. Incidentally that day I wrote two letters, one from Mr. Torr and one from Miss Hollett. Up to the beginning of this week neither had been replied to.

From your affectionate son,

* * *

Letter 59

22, Park View
Waterloo
Liverpool 22
28th October, 1941

My Dear Mum,

I am surrounded by an abundance of perfume this morning as I type this small note; the cause of the fragrance being that the landlady's daughter poured half a bottle of powerful scent over me on the previous night. Needless to say, it is a cause of great embarrassment to me.

There have been two more alerts since I dispatched my last letter to you. The same night there was a warning which lasted nearly an hour during which one place was heard. On Saturday there was another one but during the 90-minute alert there were two barrages of gunfire.

I had a busy day in the choir on Sunday, and concluded with a practice on the fine Rushworth and Dreaper Organ. This was the second occasion that I have played an organ since I have been living in Liverpool. The practice lasted 40 minutes, every minute I thoroughly enjoyed.

Last evening (before the scent episode) I was mentally at Totton, at Grandmas birthday party. I am enclosing my soiled linen of the past fortnight.

From Your Affectionate Son,

Edwin

***

Letter 60

22, Park View
Waterloo
Liverpool 22
7th November, 1941

My Dear Mum,

Again, I have to thank you for your fortnightly parcel, the oranges I have put aside for the moment, as I find an orange a pleasant addition to a frugal mid-day meal. I also thank you for the Gillette blades.

The weather has been cold until yesterday morning, when I found a much milder atmosphere, and later in the day it confirmed my prophecy that it would rain. It has rained again today.

There has been little enemy activity here during the week.

We continue to be fairly busy at the office as another arrival has brought more work. Net week, the closing office hours will be altered by half an hour. Instead of leaving at 5:00 p.m., we will finish at 4:30 p.m.

In all probability, this will be the weekly letter of my 12th term in Liverpool as next Friday should be my travelling day. Although it has been a seven-week term, it has passed like a ship in the night.

As there seems to be a dearth of news, I will close now with the anticipation of my week leave.

From your affectionate son,

Edwin

***

Letter 61

22, Park View
Waterloo
Liverpool 22
21st November, 1941

My Dear Mum,

The train to Basingstoke was not over-anxious to reach its destination before time, although it was late leaving the Central. It stopped outside Eastleigh for quite 30 minutes, and stayed at the station for some considerable time before proceeding on its journey. In fact, it was gone at ten o'clock before we reached Winchester.

Opposite to me, in the compartment were two military policemen, with a young man in overalls and trilby hat sitting between them. It was not until sometime after I discovered awkward one-handed movements of one of them that the man in the middle was handcuffed. I expect them to alight at Winchester, but they did not get off until we reached Basingstoke. There, they were going or what was wrong I did not find out. I believe they got on the same train as I did for Reading. I did not arrive at Reading until 11:30 a.m. so I did not have such a long wait on that station as usual. During the 90 minutes, I had to wait I read some columns of the Musical Times which I had with me in the case. I also had supper. The waiting room was not so full as on previous occasions when I have travelled up this way.

The time passed very quickly, and it seemed that the Birkenhead train had arrived early. It came into the station about 12:50 p.m. The carriages were not too full and I was able to find a comfortable seat. I had no difficulty in getting to sleep, even sitting upright at first, I woke up later to find one person stretched out on the other side, and that I was alone on my side. I folded my coat and used it as a pillow, and stretched out full length on the seat. I slept until the ticket inspector woke me up at Chester. In other words, I had a good eight hours sleep. It was nearly a quarter to nine then, so I decided to wash and wake myself up; there was no one else in the compartment. At Birkenhead, I had my breakfast in the Buffet and it was not until 10 o'clock that I reached the office.

Thus, I returned to commence my 12th Term in Liverpool.

The typist is still absent and I have to do all the work that she would normally do. There has been one arrival and one departure since my departure. By the way, Mr. Small left the day before I came home on leave.

Bobby has a violent cold and Mrs. Derbyshire has had one since I had mine. We played whist last night and I had two splendid handfuls of trump cards

I shall be curious to see the new clergyman at Millbrook, the next time I come home. The simulation derived from last Sundays 'Musical Festival' at Millbrook, still survives and I cannot help thinking of post-war days when new choirs will be formed to continue further musical activities.

I must close now until next week, when I hope there will be some more news to convey to you regarding my stay here.

From your affectionate son,

P.S. since I have returned, I see the British Army has come to life in Libya – this time I hope they will totally annihilate the Italian defenses and finishes the job.

* * *

Letter 62

22, Park View
Waterloo
Liverpool 22
28th November, 1941

Dear Mum,

I received your parcel yesterday morning before breakfast. I have settled down again to my North Western life, only to look forward to the next leave, and in the meantime to pass the time away by working at the office and by amusing myself at card games in the evenings.

Last Friday Mr. Mason, the organist of Christ Church called at 22, enquiring after me as I had not been in the choir for two

Sundays. I thought it very kind of him to call. I explained the reasons for my absence. I sang in the choir last Sunday, and after the morning service there was a carol practice in the choir vestry.

My first Saturday afternoon of the 12th term was spent walking. I visited a church in Waterloo with the object of seeing the organ. I did not see it as there were people about. I walked away from Waterloo in the direction of Thornton. On my return journey I met Mr. and Mrs. Carpenter going their home in a quiet avenue off the main road.

On Wednesday I entered the Cathedral for the first time since you were with me in August last. Photographers were taking pictures of the windows from various positions. There was, unfortunately, no organ interlude.

At the office we have another arrival, the first of a group, which will make us busy again. Although this week has been the quietest for many weeks. A member of the Outward Freight Department has been transferred to our Department today, to augment the number dealing with the stores side. In rank he will be next under our Chief Clerk. Our Department now consist of the following:

Mr. Child	Asst. Victualling Superintendent
Mr. Whiting	Asst. Asst. Vict. Supt.
Mr. Wolff	Shore Chef
Mr. Durkin	Chief Clerk
Mr. West	Stores Clerk
Mr. Carpenter	Assistant Stores Clerk
Mr. Rice	Assistant Stores Clerk
Mr. Knowles	Cuddy Clerk
Mr. White	Assisant Cuddy Clerk
Me	Junior Clerk
Miss Thompson	Typist

From Your Affectionate Son,

***

Letter 63

22, Park View
Waterloo
Liverpool 22
2nd December, 1941

My Dear Mum,

Under a November sky, I crossed the Mersey in the ferryboat 'Marlow' to Seacombe, last Saturday afternoon. Ships, large and small, sat tranquilly in the blue mist which had spread itself out over the river, obscuring the opposite bank and causing the outlines of objects to appear uncertain, as if they were only dark patches of mist. The journey was similar to many others that I had made last year at the same time when I used the ferry boats to cross the river each day. It was this similarity which gave me a feeling of returning to an old routine, and as though I had not in the meantime lived elsewhere. There were the same crowds of workers and shoppers pushing and shoving on and off the boats at either side of the river all was the same.

At Seacombe Pier Head, I found the same Wallasey Corporation Buses waiting for their passengers from the ferry boats and I felt quite at home there. I left the bus at Magazine Lane and walked to the park, past the desolate 'swelling' houses of Vale Drive. In passing, I remember the frightful nights that caused their desolation.

I sat on a seat in the Park in view of the bandstand. I meditated as I sat and ate cream cheese and cream crackers. Mr. Miles and I sat on the same seat and there was a crowd of people there then. There was no band there now, and no people – I was alone on the seat with my thoughts. The only sounds were there ships sirens and the hooting of tugs, which echoed mysteriously in the river valley.

I experienced a peculiar morbid enjoyment in reliving for a few brief hours the earlier days of my evacuation here. When I found I had eaten all the food, I had brought with me, I stirred myself and found my way out of the park on to the promenade. I saw all the familiar objects; the Tower, the Pier, the Battery, and the Palace. I perused the course of the promenade out of New Brighton. I had an idea I would reach Wallasey Village to see the little grey stone church which can be seen for many miles on a clear day – in fact, I have distinguished its outline from Waterloo. Having conceived the idea, I did not walk leisurely, but moved at a vigorous pace.

Owing to the mist, it was some time before I could identify the church clearly. I found the doors stood open invitingly. I searched for the organ, which I found on the right-hand side of the chancel. It is one of those boxed up affairs. However, it is a Nicholson, and should say it is a 2-manual.

In a drier and colder atmosphere on the following day, I went to Christ Church and sang in the Advent hymns. The chants were Mr. Mason's composition. For the last four years, I have spent Advent a different church.

> *Last year as a choir member Holy Trinity, Millbrook, Southampton   1938*
> *First year as an organist, St. Georges, Southampton Docks 1939*
> *First year of evacuation, choir member at St. James, New Brighton   1940*
> *Second year of evacuation, choir member at Christ Church, Waterloo   1941*

I enjoyed both services last Sunday. After the morning service, Mr. Mason took the choir over to his house in Great George Street. There we practiced carols in an upper room, which was by no means large enough to contain the large volume of sound.

I have realized this week that I have spent the greater part of this war in Liverpool. The war has been on for 27 months. I have been in Liverpool nearly 16 months.

Rice returned from his week end leave on Monday morning, after having a splendid but short time in the South. He gave me a piece of a homemade cake from Eastleigh, at lunchtime that day. We usually give each other tit-bits like that on our respective returns.

I am enclosing my soiled linen.

Mr. James, I presume has now left the Parish of Millbrook. Millbrook Church will never be the same as I have known it – Midnight Mass, Processions, Candles, Incense, and Father James!!!

From Your Affectionate Son,

Edwin

***

Letter 64

22, Park View
Waterloo
Liverpool 22
5th December, 1941

My Dear Mum,

I thank you I am getting further ahead with you for your letter which I received on my return to the digs on return from the Office last evening.

When I come home again, I shall see a new home! I suppose windows will be put in the dining room, to replace the black patches which now serve as windows. It will be good to have the house ready for Christmas.

I am getting further ahead with the Christmas Program. Today, I have completed my Christmas card Blitz. What do you and dad want for Christmas? A variety of items would assist me in choosing.

You said your goodbye to Father James. When I go home and attend Millbrook Church I shall be the oldest member of the church present. Office routine is much the same this week. Miss Thompson has returned but to a different department.

It is three weeks to Christmas and I should be able to come home, although I have not started asking yet – a little too early.

From Your Affectionate Son,

Edwin

***

Letter 65

22, Park View
Waterloo
Liverpool 22
30th December, 1941

Dear Mum,

The 13th term has begun. It opened by my return in the usual manner. It was a cold journey. The Birkenhead train did not leave the usual platform hence, it was a little time before I found a general waiting room. It was cold on the station. After two and a half hours of wait, the train roared in. It was not overcrowded and I found a seat. The handle controlling the heating of the compartment indicated that the heat was on, the atmosphere was contrary to that indication. As the journey proceeded, I dozed, woke up at the intervals to find my feet cold. I was glad that I had kept on my overcoat.

I arrived at Birkenhead at about 9:15 a.m., had breakfast and thence I went to the ferry. The fog bell rang rhythmically accompanied by flickering flares on the quay. The ferry boat 'Bidston' moved cautiously across the Mersey, and docked in Liverpool at 10:30 a.m. I did not feel like work when I arrived, I never do on the arrival morning. I was thankful to enter a warm building.

The day passed quickly and I arrived home at 6:30 p.m. There, I found the Merchant Seaman and his wife again. The next person I met was Joan who was pleased to see me. She presented me with the post which had collected during my short absence.

Among the correspondence was a card from Auntie Marie, a card from Nordic and one from John. Included in the mail was a communication from the Government. Joan was anxious about the contents it was posted on the 23rd December, and bore an official paid stamp.

I gather the contents that I have to spend a couple of days in Padgate, Warrington, Lancs. I have to report there on Friday, 9th January, 1942, to be enlisted in the ROYAL AIR FORCE VOLUNTEER RESERVE. The envelope also contained a railway warrant for the return journey from Liverpool to Padgate, and a P.O. for 4/- being one day's pay in advance. I *have shown this correspondence to my employers, who have duly noted my forthcoming absence from duty.*

The following is an extract from the letter:

"Dear Sir, in accordance with the National Service (Armed Forces) Acts, you are called upon the service in the R.A.F.V.R. and are required to present yourself on…etc.' In the last paragraph, it points out that I have to emphasize upon my employers that my absence from my present employment will be only for a day or so.

If you are writing to Bedford and you are going to thank them collectively for the Xmas presents, if so, I need not to write. I found that I have left my diary at home, but cannot recollect where I put it. If you should come across it please forward it.

From Your Affectionate Son,

***

Letter 66

22, Park View
Waterloo
Liverpool 22
2nd January, 1942

My Dear Mum,

The first week of the 13th term has neared its end, and a New Year has begun. During the week, I have indulged in a little music at times when the establishment has been left with myself as sole occupier with no one to disturb. When the house is full, I read the book of literary extracts which Mrs. Reeve kindly gave me.

Anticipation is growing within me as I look forward to my short journey to Warrington, and brief stay at that town. As I will not be at my usual post next Friday, instead of typing the weekly letter to you on that day I will endeavour to type it one day earlier. If I have time at Padgate, I will write again from there.

I hope to receive my clean laundry before I depart, I enable me to take some with me, should I be detained over the next week end.

I have instructed two people in the laws of the new demon game, Mr. Warsop and Joan. The former person was not greatly impressed by the display of cards. Joan, on the other hand received the new game with much pleasure, and she and I played for some considerable time on New Year's Eve, and again on New Year's Day.

Mr. Warren, the Merchant Seaman was much overdue from his work on New Year's Eve. His wife became anxious although, she thought he might be on the spree with the boys. I retired to bed only a little later than usual that night as I waited for supper. Mr. Warren still had not returned. I was told the next morning, he did roll in (I mean it, too), too tight for anything.

The tempo at the office is still fairly hectic, it has been after five o' clock each night when I have left the office. I am looking forward to attending Christ Church again, on Sunday, and I hope to attack the Organist regarding organ practice.

Thus, I will conclude this letter, the first of 1942.

From Your Affectionate Son,

***

Letter 67

22, Park View
Waterloo
Liverpool 22
12th January, 1941

Dear Mum,

I received your last letter on the Friday morning when I was due to leave for Padgate. I had no time to read it at breakfast. I had enquired at the Railway Station previously, how the trains ran to Padgate, and planned accordingly. I intend to travel by the 9:30 a.m. from Liverpool Central (L.N.E.R)

It was a fine morning but a cold one. I was well-armed for the excursion, with overcoat, scarf, and hat; gas mask, and case full of toilet requisites, and chocolates, and addition a parcel of laundry which I hoped to post in Liverpool. I left the house at 8:30 a.m., and an obstacle race commenced.

A little way down Park View my gas mask case gave way, compelling me to return to No. 22 with it. On opening the door, the puppy rushed out, and overwhelmed by sudden liberty sped down the road and through the railings into the park. Time was becoming a precious factor, more precious than a gas mask or puppy. The puppy bounded on through the park some way ahead of me and I decided to let it go. The animal ran on ahead for some time. Finally, I grabbed at it, but it darted off. I was in desperation. I looked back and in the distance, I could discern the portly figure of my landlady following me in pursuit of the puppy. At last, I gained speed on the animal, and gripped it round the scruff of the neck, swung myself round and raced back again. The combined speeds of the oncoming landlady and my full reverse quickly brought us together, and I relieved myself of the live burden, swung round again and headed eastwards once more at full speed.

Twenty minutes had been lost. I was on Waterloo platform at 8:50 a.m. and met Mr. Durkin there. I arrived at Exchange Station at 9:15 a.m. I took short cuts to Lord Street, then I met someone else, a young lady whose acquaintance I made in 1940. I did not allow myself to be delayed for many minutes. I was soon in the train bound for Sheffield. Within 30 minutes, I was on Warrington Station as I had to change here. The next train was for Padgate. On this train, I met 4 other fellows, bound for the same place. We arrived at the little station 10 minutes later. There was no scenery on the journey as the ground is perfectly flat around here. It was much colder here. Puddles were ice and frost was on everything. We were admitted to the RAF Station by showing our Notices. The barbed wire gates had closed behind us, and we formed up in 3's and marched off to the Intake Section. Here we were issued with a large mug, knife, fork and spoon. Here we had out snack.

I will not bother to go into the following details but to summarize, we endured long waits at various stages and processes, for tests and interviews. The meals were good and the sleeping accommodation was good.

I thoroughly enjoyed the expedition. I passed a Morse Aptitude Test and satisfied all questions asked. I left Padgate a member of the RAFVA, with a badge in my coat to certify that statement. I also have a number: 1683750. I have been taken as a wireless operator and have been placed on deferred Service for approximately six weeks.

I left Padgate at 12:30 p.m. I went through the process quicker than some, many fellows had been there 3 days.

On Saturday night, I was tired and went to bed at 10 o'clock. Joan woke me up, I was in a daze, she said that the warning had sounded and thee had seen a good deal of gunfire. I had slept so heavily for the few minutes that I had been in bed, that it some minutes before my mind realized the situation. Then gunfire pierced my ears confirming her statement. I got up and went down stairs for 2 hours until the all clear had sounded. 12 people were killed in the raid and one place was brought down. I did not rise until past 11 o'clock on Sunday, therefore I made up lost sleep.

Hoping you are all well, including the cats.

From Your Affectionate Son,

Edwin

P.S. I have not received my Health Insurance Card yet!

* * *

Letter 68

22, Park View
Waterloo
Liverpool 22
17th January, 1942

My Dear Mum,

The beginning of the week seemed quiet, after the excitement of the previous weekend at Padgate. But the tempo of things has increased towards the weekend culminating with a rush today.

I thank you for your letter which I received on Friday evening. I have dispatched my full insurance card this morning, and handed the new one in to our cash department here, where it will be duly stamped each week, until I am finally called away to the Forces.

At first, I estimated that this would be little news this week, as I wrote on Monday. Last week I wrote to Auntie Ruth to inform her of my visit to Padgate, and received a reply on Wednesday of this week. Correspondence seems to be flying about quickly; today C.A.R. said that Mrs. Willis at New Brighton has some letters for me. Letters! Letters! Letters!

There has been no snow here yet, with the exception of a few small flakes which fell for 10 minutes on Monday, although the atmosphere has been icy. Yesterday and today have not been so cold, and the prospect of a fall of snow is at an end, providing the temperature does not recede.

When the household retires to the cinema (usually twice a week) it provides me with two evenings in the week in which to practice on the piano, undisturbed. Thus, on Tuesday of this week, I made another attack on Raff's Prelude and Fugue in E Minor. The following night, we remained within doors and listened to Verdi's Opera 'Aida', and the music provided ample source for repose as well as entertainment for the greater part of the evening. We have played Option Whist three times this week. Mr. W has been a habitual achiever of the highest score.

During the week, it reached my ears that there was a film called 'the Black Cat'. I decided to see it.

I voiced my desire, and Joan wished to accompany me. On Thursday night, therefore, Mrs. D, Joan, and I went to the Plaza at Waterloo. The film was classed as a horrific picture, but it was broken by a half-wit. There was a supporting Ministry of Information film, and two other short films.

The same night Bobby returned from London, after his Christmas holidays. The night, previously, I had been shifted into a middle room, larger than the other room. Bobby and I, now occupy this room.

Mr. C. H. Moody has left Liverpool for the last time this morning. I saw him yesterday and shook hands with him. He has his call up papers, and is being sent to Tidworth to join the Royal Army Corps. He is eighteen and a half years of age, and registered in December last. He had his medical in the same months, and has now departed. Another junior of the office had his medical yesterday.

C.A.R. is going on weekend leave today, and will remain home until Thursday night.

This afternoon, I will go the New Brighton, possibly it may be the last time, as I will not have many weekends here. The main reason of my visit is to see Mrs. Willis, to collect some correspondence.

On Sunday, I will make my first appearance at Christ Church since my weekend at Padgate.

From Your Affectionate Son,

[signature]

***

Letter 69

22, Park View
Waterloo
Liverpool 22
27th January, 1942

Dear Mum,

I thank you for your parcel which I did not receive until Monday evening, 26th inst. I did not write on Friday last as I had not a letter to reply to.

On the Saturday afternoon mentioned in my last letter to you, I visited New Brighton for the purpose of collecting a letter from Mrs. Wills. As you thought, it was a letter from Coventry, to which I have not yet replied. It is a letter containing 37 words, in which she mentions that she has been promoted to the position of Secretary. She reviews her stay in Totton in July 1941, and regretted not seeing me there. I also called on Mr. Miles in Manville Road, and he requested me to play his piano. I returned to Waterloo in time for tea, after which I was commissioned to convey a young friend of Joan's, who had called, to her home in an adjacent borough 20 minutes walking distance.

I woke up late on the morning of Sunday. It had snowed a little in the night, just sufficient to cover the ground to a depth of half an inch. In the afternoon of that day, I attended Evensong at Christ Church. After the service, I attacked the organist on grounds of organ practice, and obtained permission to practice on the fine organ in the church for the nominal fee of 1/- per hour. I was very pleased with my success as a result of the 'raid' on the organist. In the evening of the same day, I again effected a similar duty to the one of the previous night.

It was cold on Monday and this I consider marked the beginning of Liverpool's snowstorm. I had lunch at another café – Lyons in Cook Street. I was accompanied with one of the company's remaining junior. After we had satisfied the appetite of our inner men, we visited the stores in Church Street. It had started to snow before we returned to the office. It snowed on through the afternoon and evening. At No.22, the next morning, we arose to find the earth covered with snow from 9" to 3 feet deep. Joan and I set off to the station at Waterloo, but we found the trains were not running. Crammed buses ploughed their way over the snow at long intervals, and defied any attempt to travel by them. We walked to Seaforth to find out whether the overhead trains were running – no. Several times, I put my foot in snow to such a depth that I rammed large quantities of it up my trouser leg. We returned to No. 22, and commenced thawing out at 11 o'clock. I found that the snow inside my trouser had solidified and had formed a thick cylinder of ice, while I was seated in the hearth by the fire!

I had dinner and set out again with Joan for our offices. The trains were working by then. We arrived at our destination at

2:00 p.m. to commence a day's work! I was glad to return home, as I felt very tired after the trying walk across the snow.

By Wednesday, the temperature had dropped to 22', but here was no more snow, and the trains were still running, but very spasmodically, causing great crowds waiting for them. That day, I reached the office at 9:30 a.m.

The temperature remained the same on Thursday and at midday, another heavy fall set in, it continued until late at night. The wind changed to N.W., and the temperature rose rapidly; the snow changed to sleet, and then rain. Although the fall of snow during the day stopped many trains, and I had to wait for over an hour at Exchange Station in a vast crowd before a train came in.

The next morning, Friday, it was raining hard on the snow; the three-foot mountain had reduced to one-foot hills. That day I arrived at the office at 10:00 a.m.

Last Saturday, there was little snow to be seen, and it stopped raining in the morning. In the afternoon, I thoroughly enjoyed myself on the organ at Christ Church, played Bach's Fugues and two Preludes by Hesse. Next Saturday, Bobby and I are attempting the Grieg Concerto on Piano and Organ. At the Organ, I shall play from the orchestral accompaniment.

To the utter disgust of the other members of the household on Saturday night, Bobby and I conversed in French for over an hour.

I have a letter from John Le Brocq; he has been called to the Forces. He is going to be a wireless telephonist in the Navy. I hope, he is sent to Tatchbury Mount [just outside Totton], in which case I could see him when he is home. He is going to write me as soon as he gets a spare moment, whereby I shall learn his new address.

I also thank you for the fruit and chocolate in the parcel.

From Your Affectionate Son,

***

Letter 70

22, Park View
Waterloo
Liverpool 22
31st January, 1942

Dear Mum,

I thank you for the socks, the parcel containing same was handed to many Mrs. D at Dinner (6:30 p.m.) on Friday evening.

By the time, I have written this letter and you should have received my parcel containing my soiled linen which was dispatched during the course of this week. It appeared to be a long time since I had heard from you, before your parcel came on Tuesday, it was 6 days on its journey to me.

I get a fair amount of news from members of the staff who periodically visit the home town, during my terms here.

Each week for several months, I have listened to the Friday night (8:30 p.m.) Tommy Handly broadcast, in which he presents himself as the Mayor, and impersonates himself as the loyal Mrs. Mopp ('can I do you now sir') and the sophisticated Claud who habitually insists on exercising his courtesy by repeating 'No, after you, Cecil.' Interruption from 'Funf' and the 'Diver', always seem to upset the Mayor's daily routine, which the Mayor usually reads from his own diary early in the program. If you have not heard this broadcast, tune in to the Forces wave length on Friday next at 8:30 p.m., and enjoy some sharp wit.

Great musical preparations have been made by Bobby and I during the course of this week. Attention has been given to the first movement of the Grieg Concerto in A minor. It is hoped to be rehearsed at Christ Church, Waterloo, this afternoon on the piano and organ. Providing the piano and organ are in tune together, the event will be a great one in the history of my musical achievements. My excitement this morning is only comparable with those occasions when I have been anticipating great festivals.

The weather has been mild after out five days of snow, although we have not been spared a drop of rain since.

Regarding the war situation, I am beginning to wonder where the German Air Force is, and what it is doing.

From Your Affectionate Son,

[signature]

***

Letter 71

22, Park View
Waterloo
Liverpool 22
7th February, 1942

My Dear Mum,

I did not remain in the office one moment longer than necessary last Saturday. By 12:30 p.m., Bobby and I were together on the piano at No. 22, putting the finishing touches to the First Movement of Grieg. We broke off activities for lunch, a break which gave the other members of the establishment time for a breather, while two musicians gorged ourselves with chips and fish cakes, boy-oh-boy! Barely did we allow out inner man to satisfy their appetites before we set ourselves again to the task of combining our musical souls on the piano. We practiced until 3:30 p.m. We, then went to the Church to meet a friend of Bobby, his name is Charles. We had an amazing afternoon of playing together all afternoon.

In the evening, Bobby and I went to the Plaza. The main film was 'This man Reuter'. It revealed the beginning and building up of the Reuter News Agency.

I attended both services at Christ Church on Sunday, the 1st February. A half-inch layer of snow had fallen during the previous night which resulted in a poor congregation. In the afternoon, I bade farewell to Mr. Burnett, the organist of Bath, as he has sailed at last from this port in an ironclad.

On Monday, I visited the Cathedral, where I was fortunate in hearing a boys choir practice, accompanied by a grand piano. The practice was taking part behind the high altar screen. In this respect, Tuesday repeated itself as I again visited the Cathedral. I thoroughly enjoyed the practices which lasted for the duration of each of my visits (30 minutes). On Wednesday, I went again but all was silent.

A little more snow fell on Monday, and while I slept in Martin's Bank Building, a good quantity fell, but by the morning it had degenerated into slush, and puddles.

The office, this week presents a respectable of quietude, and this being my sixth week, there is every reason to anticipate to the full the possibility of my appearance in the South next weekend; in which case I trust I shall arrive late on Friday night, 13th February.

I thank you very much indeed for the parcel, of linen; especially for the chocolate. I am going to enjoy the strains of 'Let the Bright Seraphim' at Christ Church on the Grand piano.

During the very cold period, a fortnight ago, I was grateful to Mr. Warsop who kindly lent me a jumper to wear under my waistcoat. I also have had a reason to thank Bobby for loaning me a disused pair of slippers, to wear, while I dry my outdoor shoes.

I trust you are keeping well during these changeable weather conditions.

From Your Affectionate Son,

**

Letter 72

22, Park View
Waterloo
Liverpool 22
9th February, 1942

Dear Mum,

I am enclosing my soiled linen with some odds and ends, which I can find no further use for herem and thus saving space in my case.

Saturday afternoon was another period of ecstasy. I first played parts of the 'Grieg' on the Grand Piano. Having thus, *wound myself up I set to work on 'Let the bright Seraphim'. I also practiced, during my three hours of practice.*

It is still quite reasonable to expect me on Friday evening about 11:00 a.m.

From Your Affectionate Son,

***

Letter 73

22, Park View
Waterloo
Liverpool 22
26th February, 1942

My Dear Mum,

I thank you very much indeed for the indoor shoes which arrived last Friday, stuffed with mince pies, and chocolate. The shoes fit well considering they are sevens.

I settled down to my 14th term, in Liverpool, only a few days after my return. The office temperature has steadily risen during the week, and I have accomplished some large typing jobs. Regarding Mr. Small, you could intimate to his wife that he may have two or three more days leave before he leaves, but this information must be regarded as entirely unofficial, but it yields substantial ground for optimism. The prospect is governed entirely by circumstance.

The weather has been frosty, and a little snow fell on Monday. Have you seen the young lady at your canteen, lately? I should be glad to earn of her musical ability, taste etc.

I made an impromptu visit to the large house of Charles Southern in Blundelsands. A maid answered my call, and was conducted in. Charles was at St. Michael's Choir practice at the time (7:00 p.m.), but his mother entertained me with conversation concerning music until 9:00 p.m. when he returned from his evenings activities. I was then escorted to another room where Charles and I fixed a date and time (Sunday 4 p.m.) for inspecting the organ at St. Michaels. From his memory, I obtained the specification of the organ. Then I enjoyed 30 minutes of his technique at the piano. He played the delightful sonata No. 17 of Mozart and selections from Chopin.

On Sunday morning, I enjoyed another service at Christ Church and afterwards Mr. Mason unknowingly co-operated with my Spring Offensive, by retaining the choir to decide that they should commence work on Stainer's Crucifixtion on Thursday. The performance will take place on Good Friday. The choirs of St. John's., St. Faith's., St. Luke's, and the congregational choir will assist our choir, the number should exceed 70 members.

Today, I raided the music shop in Houghton Street, where great damage was done. I captured a great composition of Henry Purcell's Trumpet Voluntary for the organ.

On that note, I will leave you until I hear again from the South. Hoping you, my father and the creatures are well.

From Your Affectionate Son,

* * *

Letter 74

22, Park View
Waterloo
Liverpool 22
13th March, 1942

My Dear Mum,

The events of the past seven days, since the dispatch of my last letter to you, contrast in number and character with the previous week. Last year at the same time, if you remember, I had a great deal to write about regarding the terrorism of the March Merseyside Blitz.

The Saturday afternoon of the last week was not similar to any previous Saturday. Then, I entered the church at Waterloo, at 2:25 p.m. I found Mr. Mason practicing. I sat in a rear pew and listened to a long recital of music by Cesar Franck. At p.m. I was sorry that he concluded. I took over and played Bach. During the practice, the usual intruder formed an audience. I also played parts of 'Finlandia' and the 'trumpet voluntary'. I did not return to the digs for tea until p.m.!!!

Bobby, the mediator between Charles Southern and myself, informed me that evening that he had seen Charles who said the arrangement for Sunday morning was cancelled, due to a choir practice following the service. I was not altogether disappointed as my services were required at Christ Church, for singing the anthem.

Mrs. Derbyshire remained in bed throughout Sunday, as she felt unwell. More bad news came to me on Monday as I heard that my former digs companion at New Brighton, Mr. Moody, was at a military hospital at Oxford suffering from the use of his right hand. I have prepared a two-page letter to him.

This week at the office has been increasingly busier with a hope of even greater activity in the near future.

This Thursday, we again had some snow in the morning. Rain later in the day rid us of it but this was followed by a penetrating east wind. In the evening, I attended choir practice Christ Church but the attention was taken over by a sudden attack of neuralgia which affected my left upper and lower jaws. I returned to the digs and gave the teeth a heavy dose of tobacco smoke which eased the trouble greatly. Today, the trouble has gone.

Last Saturday morning, a member of the outward freight dept. who has reached retirement age said goodbye to us. He is the fourth member of the office staff to have retired since we have been in Liverpool. Mr. Hume has stepped up another rung of the ladder.

Mr. Stacey, the last remaining junior of that department who I mentioned in my last letter bid goodbye to colleagues on Wednesday afternoon. He is now enjoying six days at home before going to Skegness.

At the digs too, there have been many changes this week. For about a month, Mr. Warsop has been drafted to another government department Aigburth, Liverpool, 19. This has necessitated a 1½ hour journey across the city each day. At last, he has found suitable digs, and bade us farewell last Monday night. No. 22 to me, will not be quite the same without him, as he has been resident here since before August of last year.

Mr. P Burk, a man who has only been at No. 22 for about a month, has received instructions to proceed abroad. He is working at present in a government hospital, where he deals with pharmacy. Thus, he will depart next week. He has played chess with me during the week.

Everyone's leaving so it seems. Now, I have put all this news first to make sure you would read it, for the next paragraphs, the most important, if they had been written earlier in this epistle, may have prevented you from reading further.

On Wednesday of this week, on my arrival at the digs in the evening, I received my call up papers instructing me to report to Padgate on the 30th march. A travelling warrant has been enclosed, not giving the name of the station from which I depart, therefore it will serve to take me to Padgate from anywhere in Great Britain.

The first move I made on Thursday morning was to inform Mr. Durkin, my chief Clerk, of my call up. He instructed me to type a memorandum to our general department to this effect. The majority of the day I was dealing with the technical problems of the case.

Today, I took further steps in negotiating my leave. From today, it is 17 days before I have to report to Padgate. As yet, it is possible that I shall travel home towards the end of next week, probably Thursday 19th inst. This date is entirely unofficial.

Joan seems to regret the probability of my early departure for she clings around like a leach.

I will forward my parcel of soiled laundry on Monday in the usual manner, together with a few odds and ends to save carrying.

I will bid you farewell, for now, until I write probably my last letter from Liverpool next week.

From Your Affectionate Son,

[signature]

***

Letter 75

22, Park View
Waterloo
Liverpool 22
16th March, 1942

My Dear Mum,

I thank you for your parcel which I received on Saturday, 14th march, at 2:00 p.m., containing the laundry, chocolate, and the wonderful anthem, 'Hear my Prayer', which I played assiduously on the evening of that day.

I am worried about Auntie Marie, but as I may return to the South before another communication reaches me regarding the subject, I will have to wait until I can see for myself.

The week starting from Friday and ending on Thursday will be my last seven days in Liverpool. I have drawn up a plan for a farewell week to places and people which have become familiar to me during my nineteen months' stay in the North West.

Friday 13th March	*Last Firewatch*
Saturday 14th March	*Last organ practice at Christ Church*
Sunday 15th March	*Farewell to Liverpool Cathedral and its great organ.*
	Farewell to Christchurch, Waterloo, Mr. Mason, and members of the adult choir present.
Monday 16th March}	

Tuesday 17th March}	*Farewell to local haunts*
Wednesday 18th March}	
Departure day	*Farewell to digs and personnel*
Friday	*Farewell to Martins Bank Building*
20th March	*Farewell to Colleagues*
	Farewell to Lime Street Station

FAREWELL TO LIVERPOOL

Already some of these plans have come to pass, others will be done so after this last letter from Liverpool has left me, until the last item has been fulfilled, and I am at last on the famous Lime St. – Euston train bound South.

The plans fulfilled to date I have described below. On the evening of Friday, I went back 'home' for my dinner, as the margin of daylight before blackout allowed for this maneuver. Joan's girl friend called and I engaged her in conversation, resulting in an appointment for the Saturday afternoon organ practice. My nocturnal visit at M.B.B. (fire watch) was uneventful.

Saturday was a gloriously sunny day (for Liverpool). It effected my appointment, at a specified place and time. The organ practice considered of music by Bach, Purcell, and Sibelius. From the hymn book, I selected the last piece to play on that organ. I chose the grand hymn 745 (lift high the cross).

The farewell activities on Sunday were concerned with the closing of my North Western ecclesiastical musical life. That life opened at Liverpool Cathedral on 18th August 1940, it was fitting therefore that it should be closed there. March `5th 1942, I attended a service for the last time in the great cathedral. It was a sunny morning. The service was rendered in the supreme style peculiar to the cathedral. The great organ did not assert itself to a great extent during the service, not even in the hymn 'Onward Christian Soldiers'. It was after the service that I realized I was listening to a cathedral organ. Mr. Goss-Custard, extemporized, building up all the resources of the instrument to a great crescendo to full organ, yes, ORGANO PLEINO, plus!

At Christ Church, the parting was more painful. I have been happy in the choir work, and recently at the organ. Mr. Mason was greatly upset at the final parting, not only because he regretted that I should also be a member of the choir assisting in the performance of the ordinary services. He informed me that should I at any time return to Liverpool, there would be a vacant space home address which I could fill. He asked for my home address which I willingly gave him. I said goodbye to nine of the choir men and several of the ladies, present.

Under separate cover, I have dispatched a large parcel of Laundry, scrap paper, one pair of shoes, and a dictionary.

Here ends my Liverpool adventure. From Your affectionate son,

[1] S. Nicholas Church Organ, Liverpool, specifications

Great Organ choir Organ

Contra Gamba	16	Viol di Gamba	8
Open Diapason No. 1	8	Dulciana	8
Open Diapason No. 2	8	Lieblich Flute	8
Open Diapason No. 3	8	Lieblich Flute	4
Claribel Flute	4	Harmonic Piccolo	2
Principal	4	Corno di Bassetto	8
Twelfth	2	Tromba on Choir	8
Fifteenth	2		
Tromba	8	**PEDAL ORGAN** Open Diapason	16

<u>SWELL ORGAN</u>		Bourdon		16
Open Diapason	8	Bass Flute		8
Rohrflute	8	Trombone		16
Salicional	8			
Viole	8	<u>COUPLERS</u>		
D'orchestre Vox Angelica	8	Great to Pedals	Choir	Sub Octave
Principal	4	Swell to Pedals	Swell Off	Unison
Wald Flute Echo	4	Choir to Pedals	Swell	Octave
Cornet	3 ranks	Swell to Great	Swell	Sub
Contra Oboe	16	Swell to Choir	Octave Choir to Great	
Oboe	8			
Cornopean	8	red indicates swell reeds		
Vox Humana	8			
Clarion	4			
Tremulant				

[ii] Saturday 16.11.40
ORGANIST. – H. EDWIN. HENNING ex Southampton

Program

1. Allegro Marziale — Greenhill
2. Chorale Prelude (S.Peter) — Darke
3. Festal Offertorium — Fletcher
4. Allegro Grandioso — Blair
5. Postlude in G — Wadely
6. (a) Prelude in C
   (b) Prelude in Em — Bach
7. (a) Prelude in F, No. 1
   (b) Prelude in G, No. 2
   (c) ""D minor, no. 6 — Stanford
8. Three miniatures from 12 by (sightread) — Higgs
   (a) in D, no. 1 (b) in eb, No. 7 (c) in F No. 11
9. Moderato — N.Gade
10. Overture Op. 125 — G. Merkel
11. A number of Advent and Christmas Hymns
    Nos. 45 & 60 12
12. Prelude in C Major — Bach
13. Romance in D flat — Lemare

oOo

Time taken: approx. 2 hours

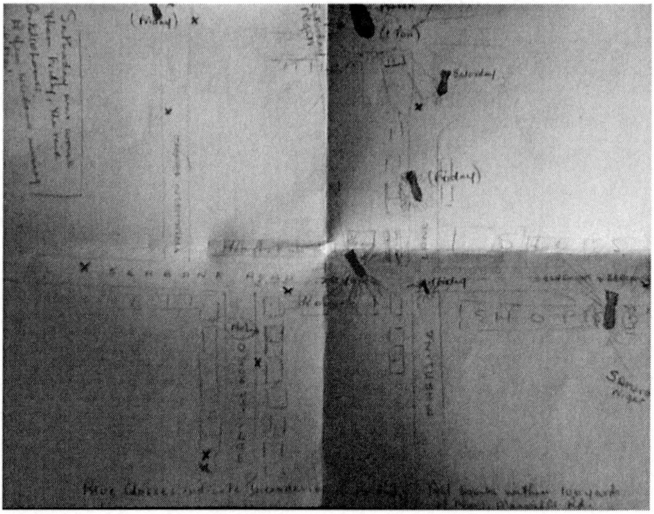

About the Editor

She was brought up and lives in Hampshire, England. After Maureen married, she pursued a teaching career. She now combines teaching with lecturing. Recently, she completed an MA in Theology in Christian Liturgy, researching the changing liturgical needs in the trenches of World War I. She is married with three children and two grandchildren. She belongs to two choirs and loves gardening.